Chinese Language Learning Sciences

This book series investigates several critical issues embedded in fundamental, technical, and applied research in the field of Chinese as second language (CSL) learning and teaching, including learning mechanism in the brain, technology application for teaching, learning and assessment. The book series discusses these issues from the perspectives of science (evidence-based approach) and technology. The studies in the book series use the methods from the fields of linguistics (such as corpus linguistics and computational linguistics), psychological and behavioural sciences (such as experimental design and statistical analyses), informational technology (such as information retrieval and natural language processing) and brain sciences (such as neuroimaging and neurolinguistics). The book series generally covers three main interdisciplinary themes: (1) fundamental investigation of Chinese as a first or second language acquisition, (2) development in Chinese language learning technology, and (3) applied research on Chinese language education.

More specifically, the book series involves seven research topics:

- language transfer mechanism in Chinese as a second language
- factors of Chinese as a second language acquisition in childhood
- cultural influence on Chinese acquisition
- information technology, corpus
- teaching material design
- teaching strategies and teacher training
- learning models
- assessment methods

All proposals will be sent out for external double-blind review. Review reports will be shared with proposers and their revisions will be further taken into consideration. The completed manuscript will be reviewed by the Series Editors and editorial advisors to ensure the quality of the book and also seek external review in order to ensure quality before formal publication.

Please contact Alice Xie (e-mail: alice.xie@springernature.com) for submitting book proposals for this series.

Denise Hsien Wu
Editor

Cognitive and Neural Foundations of Chinese Reading

From Learning to Advanced Processing and Beyond

Springer

Editor
Denise Hsien Wu
Institute of Cognitive Neuroscience
National Central University
Taoyuan, Taiwan

ISSN 2520-1719 ISSN 2520-1727 (electronic)
Chinese Language Learning Sciences
ISBN 978-981-96-6668-3 ISBN 978-981-96-6669-0 (eBook)
https://doi.org/10.1007/978-981-96-6669-0

This Springer imprint is published by the registered company Springer Nature Singapore Pte Ltd.
The registered company address is: 152 Beach Road, #21-01/04 Gateway East, Singapore 189721, Singapore

Contents

Part I
Learning to Read Chinese

Chapter 1
Exploring Chinese Reading Development: Framework and Cognitive Factors

Jun Ren Lee

Abstract This article delves into the challenges faced by children in third grade and below and those with severe dyslexia when learning Chinese. Previous literature has highlighted the significance of phonetic radicals in Chinese characters, often drawing comparisons with the regularity and consistency effects observed in English. However, upon closer examination of the Chinese characters learned by children before the third grade, it becomes apparent that the regularity of phonetic radicals alone is insufficient for accurate pronunciation, and children may not necessarily rely on them to derive the pronunciation of the entire character. The discussion emphasizes the structural characteristics of Chinese characters learned before the third grade, potential biases in character recognition assessments, and the cognitive factors that demand attention. Furthermore, the article advocates for a focus on research about teaching interventions and the optimal timing for developing various cognitive skills in promoting reading development, including phonological awareness, orthographic awareness, and morphological awareness, which impact different aspects of reading proficiency.

Keywords Consistency · Phonological awareness · Reading development · Regularity

Ai (1949) pioneered the scientific and systematic study of Chinese character learning with his seminal work, "Hanziwenti (漢字問題)." This groundbreaking book explored fundamental topics, including the structure of Chinese characters, the impact of character structure on recognition, the challenges of character recognition, children's capacity for character recognition, vocabulary acquisition, and writing skills. Many of the issues addressed in this early work continue to shape contemporary discourse in the field.

J. R. Lee (✉)
Department of Educational Psychology and Counseling, National Taiwan Normal University, Taipei, Taiwan
e-mail: jrlee@ntnu.edu.tw

D. H. Wu (ed.), *Cognitive and Neural Foundations of Chinese Reading*, Chinese Language Learning Sciences, https://doi.org/10.1007/978-981-96-6669-0_1

In the past three decades, numerous research papers have further investigated aspects such as the structure of Chinese characters, the process of character acquisition, factors influencing character recognition, and Chinese dyslexia (Chung & Ho, 2010; Lee et al., 2024; McBride & Wang, 2015; McBride et al., 2018; McBride-Chang, 2016; Perfetti et al., 2013; Reichle & Yu, 2018; Ruan et al., 2018; Zhang et al., 2023). This article distinguishes itself by focusing specifically on beginning readers (G1–G3) in the early stages of literacy development (Chall, 1983) and on severe dyslexic children who struggle with recognizing Chinese characters and have limited character recognition abilities. While emphasizing the need for further clarification and enhancement of research issues, the article refrains from providing exhaustive descriptions of relevant content and empirical findings, instead directing interested readers to consult related research papers.

1.1 Characteristics of Chinese Characters

Chinese characters were traditionally viewed as a purely logographic or ideographic writing system, where the shape of a character directly represented its semantic meaning (DeFrancis, 1984). However, understanding character structure evolved with the introduction of semantic and phonetic compounds, altering perceptions of Chinese characters (DeFrancis, 1984). Chinese characters can generally be classified into six types based on their formation: pictograms (象形), simple indicatives (指事), compound indicatives (會意), semantic-phonetic compounds (形聲), borrowed characters (假借), and derived characters (轉注). The majority of characters fall into the first four categories (DeFrancis, 1984, 1989).

The earliest known Chinese characters on oracle bones dating back to around 1000 BC reveal that 23% were pictographic, 2% were simple indicatives, 41% were compound indicatives, and 34% were semantic-phonetic compounds (DeFrancis, 1984). Pictograms are not predominant among identified characters on oracle bones (DeFrancis, 1984). Throughout history, the prevalence of semantic-phonetic compounds has significantly increased. Xu Shen's Shuowenjiezi (說文解字) (1965), compiled around 200 AD, categorized characters based on semantic radicals, encompassing 514 radicals and a total of 9,353 characters, with semantic-phonetic compounds constituting 82% (DeFrancis, 1984). Nowadays, newspapers and magazines typically utilize around 5,500 characters, with semantic-phonetic compounds representing approximately 80% (Chen et al., 2011; Zhou, 1978). Thus, the majority of Chinese characters are semantic-phonetic compounds, suggesting that Chinese reading involves phonological processing rather than direct mapping of orthography to semantic meanings (Frost, 2012; Lee et al., 2005; Zhou & Perfetti, 2023).

According to statistics from Taiwan's National Academy of Educational Research, the top 1,000 most frequently used characters, based on frequency in contemporary texts, constitute 89.2% of usage frequency, while the top 2,600 characters encompass 99% of frequency (Wu et al., 2018). Functional literacy in Chinese typically requires recognition of 2,300–3,000 characters (Huang, 1994). Primary school textbooks in mainland China, Taiwan, and Hong Kong commonly introduce approximately 2,300–2,600 characters (Hsiang et al., 2021). This data suggests that recognizing around 2,600 characters enables sounding out nearly 99% of daily life characters.

As noted in several research papers, semantic-phonetic compounds constitute approximately 80% of characters commonly used in modern daily life (Zhou, 1978). The significance of semantic radicals and phonetic awareness in Chinese reading development has been underscored in the literature (Ho et al., 2003; Li et al., 2018, 2021; Shu & Anderson, 1997; Shu et al., 2000).

The emphasis on phonetic radicals stems largely from the traditional dual-route reading model (Coltheart, 1985; Coltheart et al., 2001), positing two word identification routes: address and assembly. This model also delineated two types of dyslexia: phonological and surface dyslexia (Castles & Coltheart, 1993). The assembly route involves grapheme segmentation, phoneme-grapheme correspondence mapping, and phoneme synthesis for word pronunciation. This underscores the decoding function in alphabetic languages. In Chinese, phonetic radicals have been likened to the sounding-out function in English, and discussions have ensued regarding the regularity and consistency effects based on phonetic radical properties (Lee et al., 2005; Zhou & Perfetti, 2021).

However, does the prevalence of semantic-phonetic compounds in Chinese provide adequate phonological information for beginning readers, and can a similar regularity and consistency effect be observed in novice readers? Here, I present an alternative perspective diverging from previous research. I posit that excessively emphasizing teaching phonetic and semantic radicals, particularly for learners below the third grade, may not be the most efficacious approach. This viewpoint contrasts with the predominant focus of existing research. It is essential to clarify that this does not diminish the importance of such instruction but underscores the significance of timing. Effective instruction on character structure and phonetic and semantic radicals' roles should occur when children have acquired a certain number of characters or radicals and have reached a cognitive stage where they can employ metacognitive strategies. For children just initiating their Chinese character learning journey and those with severe dyslexia, introducing the roles of phonetic radicals may not be advantageous. This is because most characters encountered by beginners are high-frequency, yet they often exhibit low regularity in pronunciation. Consequently, phonetic radicals may not consistently provide the correct pronunciation of characters.

1.2 Characteristics of Chinese Characters for Beginning Readers

For beginning readers in the learning-to-read stage (G1–G3) (Chall, 1983), the accuracy of phonological information provided by phonetic compounds is low. In China, primary school textbooks from grades one to three typically introduce approximately

1,700 characters, with phonetic-semantic compounds constituting 65.4%. Among these, instances where the pronunciation of the radical matches that of the full character (e.g., 清, 青) and characters with different tones but identical syllables (e.g., 請, 青) account for only 24.9% of all characters (Shu et al., 2003). Consequently, less than one-fourth of characters offer phonological cues, rendering reliance on phonetic radicals for pronunciation ineffective for beginning learners. It becomes essential for learners to recognize a sufficient number of semantic and phonetic radicals to apply analogy rules for inferring the pronunciations of semantic-phonetic compounds. The critical inquiry revolves around determining when a child has acquired enough characters to effectively employ semantic/phonetic analogy.

Given the above, several issues warrant discussion: How do learners pronounce accurately when characters lack stable and precise phonetic information? How do regularity and consistency effects manifest in adult word recognition experiments? What are the similarities and disparities in cognitive factors influencing reading development between Chinese and English beginning readers?

1.3 The Locus of Phonological Information for Chinese Beginning Readers

The acquisition of accurate character pronunciation becomes pivotal when a Chinese character fails to convey meaningful phonological information. In English, phonics serve as the foundational skill for word decoding. Is there an analogous function in Chinese akin to English phonics?

Despite Chinese characters lacking inherent phonological information, it is inferred that Hanyu Pinyin in mainland China and Zhuyin Fuhao in Taiwan fulfill a phonics-like role akin to English. Both systems operate under the same principles, differing only in their notation methods. Hanyu Pinyin utilizes Latin letters, while Zhuyin Fuhao employs components of Chinese characters. Both systems facilitate sounding out for a character. The historical lineage of Zhuyin Fuhao predates that of Hanyu Pinyin, originating in the 1910s, based on the "fanqie" (反切) system, where two characters indicate the pronunciation of one character. Fanqie was used as early as the Han (漢) dynasty to mark the pronunciation of a character, with two characters indicating the pronunciation of one character. For example, 冬 is marked as 都宗切 (Wang, 1996). To simplify and standardize character marking, simplified radicals were introduced in the late Qing (清) dynasty, such as ㄅ for 包 and ㄉ for 刀.

In reading instruction, the Taiwanese government mandates the inclusion of Zhuyin Fuhao markings in textbooks for students below the third grade, with similar practices in mainland China. Thus, for beginners, accurate recognition and synthesis of Zhuyin Fuhao symbols ensure precise pronunciation of corresponding characters, bypassing the need for structural analysis. In contrast, the case of Hong Kong historically favored methods such as "look and say," but recent developments have seen the integration of Hanyu Pinyin and systematic teaching of character structure and radicals (Hsiang et al., 2021). Such emphasis on semantic and phonetic radical instruction is less common in Taiwan and mainland China.

1.4 The Measurement of Character/Word Identification

In English, the primary assessment method for word recognition is sounding out, wherein participants are presented with real words or pseudo-words conforming to phonotactic rules in English and asked to sound them out (Snow et al., 1998). As homophones are relatively uncommon in English, the ability to sound out a string of letters typically enables individuals to associate it with a sound in their mental lexicon, potentially linking it to the word's meaning. However, homophones complicate matters in Chinese, as pronouncing a word does not necessarily entail comprehending its meaning. Dependency on orthography persists; for example, 張 (zhang), 章 (zhang), and 樟 (zhang) all share the same pronunciation but possess different orthographies and meanings. Moreover, the existence of polyphonic characters often necessitates consideration of compound word components to discern meaning, such as 著火 (zhaohuo, on fire), 有著 (youzhe, to have), and 著作 (zhuzuo, publication). Consequently, relying solely on sounding out single characters may not be an effective assessment method in Chinese despite its widespread use in research.

Homophones in Chinese characters present a potential challenge for the commonly employed sounding-out method for single character measurement: recognition errors leading to correct pronunciation. Naming accuracy for regular characters may result from characters sharing the same phonetic component, yielding identical pronunciations when learned phonetically; such distinctions are not discernible in sounding-out tasks. However, simultaneous utilization of sounding out and word-generating methods allows for differentiation; word-generating tasks necessitate the generation of a word incorporating the target character. This approach accurately determines whether participants produce phonologically regular responses or recognition errors.

1.5 Phonological Regularity and Consistency in Chinese Beginning Readers

In adult Chinese character recognition studies, researchers have identified regularity and consistency effects, highlighting the phonological significance of phonetic components (Lee et al., 2005). However, in English, these effects stem from learning grapheme-to-phoneme rules and mastering rhyme units, which are crucial for sounding out.

The prerequisite for regularity and consistency lies in the repetition of basic components and sufficient uniformity in pronunciation. This poses challenges for beginning Chinese readers. Although approximately 70% of characters in the learning-to-read stage (G1–G3) are phonetic-semantic compounds, only 24% provide useful syllable information. Here, usefulness is defined as sharing the same syllable, irrespective of tone differences. However, the tone in Chinese characters serves a semantic function; characters with different tones typically convey different meanings. Importantly, the ratio of phonetic-semantic compounds to phonetic radicals averaged about 1.8 (recalculated from Tzeng & Lee, 2012), indicating that most characters encountered by beginning readers lack regularity, with fewer than two characters sharing the same phonetic component. This suggests that establishing consistency effects may be challenging.

Tzeng and Lee (2012) observed regularity effects (defined by identical syllable sounds, ignoring tone differences) in grades four to six, with consistency effects emerging only by grade five. This implies that recognizing approximately 3,000 characters—roughly the number recognized by a fifth-grade student—is necessary to manifest consistency effects involving phonetic components.

Regarding regularity, Ho and Ma (1999) observed better performance for both average and dyslexic readers with regular characters compared to irregular ones. However, dyslexic individuals exhibited similar performance levels for high- and low-frequency irregular characters. Notably, the error patterns of the dyslexic group showed fewer phonetic-related errors than the controls, suggesting that the phonetic radical may not be a primary factor in character identification for beginning or dyslexic readers. Furthermore, most errors made by participants were "do not know" errors; the high percentage of radical-related errors was calculated by excluding these "do not know" errors.

Ho and colleagues (2006) discovered that dyslexic groups learned regular words similarly to reading age control groups but struggled more with irregular words. Since beginners and dyslexic children primarily learn high-frequency characters, the regularity of high-frequency characters may be poorer than that of low-frequency characters. Overemphasizing the regularity of phonetic radicals may be misleading.

In a training study, Chen et al. (2016) found that teaching characters with the same phonetic radical to improve character recognition abilities required two conditions to be met: children must recognize more than 1,200 Chinese characters and be older than third grade. These studies illustrate the importance of acquiring a certain number of characters as building blocks and developing metalinguistic abilities before phonetic radical training can be effective.

The phonetic radical does not provide useful syllable information, and beginning readers must first learn a certain number of characters before phonetic radical training can be effective. This suggests that the role of phonetic components in developing Chinese character recognition may be less pronounced for beginning readers and those with dyslexia than previously emphasized (Ho & Ma, 1999; Ho et al., 2006; Shu et al., 2000; Tzeng et al., 2022). However, this does not diminish the importance of phonetic radical training; rather, the timing of its introduction is key.

1.6 The Number of Basic Units

If the function of phonetic radicals should be introduced later, what other factors are important for Chinese beginning readers? I argue for the importance of paired association learning between visual symbols and their sounds.

There are 26 letters and 44 phonemes in English, resulting in approximately 250 frequent graphemes and their corresponding sounds. The number of basic symbols and corresponding sounds to be memorized in English is around 276. In Chinese, even third graders must learn approximately 400 phonetic radicals and 144 semantic radicals, totaling over 800 symbols in elementary school (Tzeng & Lee, 2012). Learning these symbols relies solely on visual and auditory paired-association learning. Therefore, the influence of visual and auditory paired learning should be greater in Chinese reading development than in English. However, no research directly compares the weight of visual and auditory paired learning between Chinese and English. Visual and auditory paired-association learning may be a key deficit for severe dyslexic children in Chinese.

1.7 The Cognitive Factors Influencing Chinese Character Recognition

Peng et al. (2017) conducted a meta-analysis examining various cognitive factors' performance among groups with reading disabilities, age-matched controls, and controls matched for reading age. Compared to age-matched controls, children with reading disabilities exhibited lower performance across nearly all assessed variables to varying extents. Reading disabilities were associated with moderate deficits in rapid naming and mild deficits in orthographic knowledge compared to controls matched for reading age. However, the study acknowledged limitations related to potential comorbidities and the control of IQ.

When exploring the impact of cognitive factors on reading development, it is crucial to consider how specific tasks may shape observed outcomes (Peng et al., 2017; Zhang et al., 2023). Subsequently, the operational definitions of variables and their potential influence on results will be addressed.

1.8 Visual Processing

The square-shaped nature of Chinese characters prompts discussions about visual processing's influence on Chinese reading skills. Early evidence suggests a significant correlation between visual processing and reading development in Chinese, a link not observed in English (Huang & Hanley, 1995). However, this correlation lacks consistent confirmation (McBride-Chang & Kail, 2001; Siok & Fletcher, 2001).

Studies on visual processing often utilize visual-verbal association or visual-orthographic tasks, introducing confounding variables. Ho et al. (2002, 2004) demonstrated that when controlling for reading level or accounting for variables such as age, IQ, phonological awareness, phonological memory, orthographic processing, and rapid naming, the effects of visual discrimination and visual memory vanish entirely. Yang et al. (2013) conducted a meta-analysis focusing solely on visual perception, speed, and memory, finding a correlation coefficient of approximately 0.35 between visual processing and character recognition ability. However, this correlation diminishes significantly for participants beyond the first grade.

Thus, visual processing plays a role in Chinese reading, primarily during the early reading stages. Once a certain level of character recognition ability is achieved and radicals are learned, structured character recognition becomes more effective, diminishing reliance on visual processing.

1.9 Phonological Awareness

The phonological awareness serves the speech sound segmentation function and might further re-represent speech sounds using phonological coding symbols. Phonological code's introduction enhances phonological awareness (Morais et al., 1986), and phonological awareness enhances the connection between a writing symbol and its sound. In this scenario, phonological awareness is not directly related to the mapping between orthography and phonology in a writing system, which Chinese is an example. By taking phoneme tapping as an example, which asks the participant to use finger tapping according to the number of sounds for a word, "cat" as a stimulus target shall be tapped three times (Liberman et al., 1974). It is involved in segmenting speech sounds but is not relevant to orthography. The opaque relationship between orthography and phonology in Chinese is irrelevant to the impact. However, the phonological coding symbols influence speech sound representation and affect phonological awareness performance. Children in Hong Kong who were instructed through the look-and-say method without a phonetic coding system their performance in phonological awareness than children from Taiwan and Mainland China (Cheung et al., 2001; Huang & Hanley, 1995; McBride-Chang et al., 2004).

The operational definitions of phonological awareness include segmentation, deletion, blending, oddity tasks, spoonerism, and phoneme reversal (Adams, 1994; Landerl et al., 1997; Paulesu et al., 2001). Different phonological awareness tasks have varied cognitive demands and correlations with reading (Stanovich et al., 1984; Yopp, 1988).

Research on Chinese phonological awareness encounters two challenges. The first is the low reliability of preschool phonological awareness tasks, and the second is the lack of research establishing causal relationships between phonological awareness and Chinese literacy development.

In preschool, deletion and forced-choice tasks are commonly used as phonological awareness tasks. Preschool children can only perform syllable deletion tasks when using deletion tasks, while sub-syllable deletions are beyond their ability (Cheung et al., 2001; Shu et al., 2008). When employing forced-choice tasks, researchers often calculate whether the performance exceeds the chance level using group statistics rather than considering individual participants' performance against guessing probabilities (Lonigan et al., 1998). Current research suggests that while group-level accuracy may surpass chance values in forced-choice tasks, only a few participants exceed guessing probabilities when assessed individually (Ho & Bryant, 1997a, 1997b; Shu et al., 2008). This is also shown in the reliability. Much research reported that reliability coefficients for preschool children are generally below 0.5, falling short of the recommended threshold (Ho & Bryant, 1997a; Ho et al., 2002; Shu et al., 2008).

Early discussions on the influence of phonological awareness on Chinese literacy yield varying perspectives. Some researchers argue that phonological awareness has no effect, while others claim it has an effect similar to that in English. This issue can be divided into two arguments: first, whether phonological awareness affects literacy development, and second, whether the impact of Chinese phonological awareness is comparable to that in English. Meta-analyses have shown a correlation between phonological awareness and Chinese literacy, but the correlation is lower than in English (Ruan et al., 2018; Song et al., 2016).

An important point is that meta-analyses only provide correlations and cannot establish causation. While studies on English phonological awareness have established a causal relationship between phonological awareness and reading development through cross-sectional, longitudinal, and intervention studies, Chinese research lacks intervention studies. Both cross-sectional and longitudinal studies in China and Taiwan support the function of phonological awareness, but intervention studies show that enhancing phonological awareness skills does not affect word recognition (Zhou et al., 2012). The only intervention study demonstrating an impact on literacy development through Chinese phonological awareness teaching occurred in Hong Kong, albeit with a low effect size (Wang et al., 2017).

1.10 Morphological Awareness

McBride-Chang et al. (2003) defined morphological awareness as the comprehension of words' internal structure and significance, encompassing derivational and inflectional morphology. Derivational morphology involves prefix morphemes (such as "un-"), suffix morphemes ("-ation"), and compound words (e.g., "cowboy"). In contrast, inflectional morphology pertains to grammatical alterations in words (such as plural "s" or past tense "ed").

In Chinese, each character represents a morpheme, with most words being disyllabic. Thus, compound words in English serve as a more fitting analogy for Chinese morphology. For example, in English, "cow and boy" amalgamate to form "cowboy," whereas in Chinese, the first character of "電鍋" (electric cooker) is "電," which means "electricity" in English. The second character is "鍋," which means "pan" in English (Liu & McBride-Chang, 2010).

However, in English, single morphemes constitute 45%, while compound words comprise only 25% (Dupuy, 1974). In contrast, bimorphemic compounds account for 73.6% of Chinese (Ma, 2014).

Past research has revealed a correlation between morpheme awareness, vocabulary skills, and reading comprehension. Given that most words in Chinese are disyllabic, in contrast, English words typically consist of a single morpheme, it is reasonable to infer that morphological awareness holds greater significance in Chinese, as there are relatively fewer compound words in English (Liu & McBride-Chang, 2010; Shu et al., 2006).

When examining morphological awareness, it is essential to control for intelligence, especially the assessment of reasoning ability, as morpheme awareness tests involve reasoning processes. Eliminating the influence of reasoning ability could better underscore its linguistic value.

Another issue is that past morpheme awareness tests have predominantly focused on the operational rule of transparent words, with less emphasis on opaque words like "天花" (smallpox), which has nothing to do with the first character "天" (sky) and the second one "花" (flower), necessitating further research on how learners acquire opaque words.

Regarding the developmental timeline, words are the fundamental learning units rather than characters. Learners initially acquire oral vocabulary; then, through literacy, they establish the meaning of characters based on their accumulation within words. Subsequently, through the meanings of characters, they can derive the meanings of new words. Therefore, morphological awareness likely emerges later than phonological awareness, probably after a certain level of character recognition (Tong & McBride-Chang, 2010).

1.11 Rapid Automatic Naming (RAN)

Rapid Automatic Naming (RAN) refers to participants' ability to swiftly name a series of numbers, letters, or pictures, primarily involving efficiently extracting highly familiar information. Numerous studies on Chinese reading development underscore the importance of rapid automatic naming, suggesting its significance surpasses that of phonological awareness (Chang et al., 2014; Ho et al., 2004; Tan et al., 2005; Tzeng, 2005). For instance, Tzeng (2005) conducted longitudinal experiments on rapid automatic naming in Taiwan, revealing that early rapid naming ability effectively predicts later reading comprehension. Researchers in Hong Kong suggest that the most prevalent subtype of Chinese dyslexia is associated with rapid automatic naming rather than phonological awareness (Ho et al., 2004). Tan et al. (2005) found no correlation between phonological awareness tests and character recognition after controlling for non-verbal IQ, while significant correlations were observed between automatic naming and character recognition for beginning readers. Indeed, Song et al. (2016) found that RAN correlated significantly with reading accuracy and reading fluency, with RAN correlating more strongly with reading fluency than reading accuracy. Therefore, rapid automatic naming is considered an indispensable component of cognitive testing in Chinese reading development.

However, limited discussion has been on what rapid automatic naming precisely measures. Research indicates that rapid automatic naming correlates with accuracy and fluency in word recognition and comprehension. Nevertheless, naming non-consecutive single items (discrete trial naming) shows no effect (Logan et al., 2011). Rapid automatic naming represents the extraction and pronunciation of familiar symbols. If the effect of single items is significantly lower than that of serial rapid naming, sequential processing may be more critical than retrieving individual symbols. Indeed, the pause time in rapid automatic naming exhibits higher predictive power for reading fluency than articulation time, although it does not impact spelling and reading comprehension.

Similar assessments of rapid automatic naming tests are used across Mainland China, Taiwan, and Hong Kong. However, previous studies in these regions indicate that children from Taipei (16 s) are faster than those from Beijing (18 s), and children from Beijing are faster than those from Hong Kong (23 s) (Pan et al., 2024). The reason for the slow response by Hong Kong children is unknown. This variation may result from differences in the pronunciation of numbers in these regions. Due to varying syllable lengths in number pronunciations, differences in the speed of rapid number naming may arise. Such differences may not significantly impact regression analysis within each region's data. However, if direct comparisons are made, careful examination of the language difference is important.

1.12 Phonological Working Memory

Phonological working memory refers to the storage and manipulation of phonological information. One common assessment of phonological working memory is the pseudoword repetition task, which requires participants to repeat nonsensical speech sounds. The use of pseudowords helps eliminate interference from semantic information. Previous research has demonstrated that phonological working memory impacts word recognition, vocabulary, and comprehension in English (Carretti et al., 2009; Gathercole & Baddeley, 1993). However, some studies suggest that phonological working memory only affects reading comprehension, not accuracy and fluency (Oakhill & Cain, 2012). Melby-Lervåg et al. (2012) conducted a meta-analysis and found a correlation of 0.34 between phonological working memory and word/nonword decoding. When compared with peers of the same age, the effect size is 0.71, but when compared with peers of the same reading age, the effect size is −0.09. However, when statistically controlling for phonemic and rime awareness, phonological working memory does not impact decoding.

Although phonological working memory has a limited direct impact on word recognition, it may influence word recognition through phonological awareness. This is especially noteworthy as phonemic awareness has various operational definitions, and the demands on phonological working memory differ across phonological awareness tasks. When tasks require more executive functioning, does this strengthen the influence of working memory? Another consideration is whether improving reading skills enhances phonological working memory. After all, familiarity with symbols can enhance phonological representation efficiency, thereby increasing phonological working memory capacity.

Oakhill and Kyle (2000) found that, after accounting for reading ability, the sound category task in phonological awareness tasks is influenced by phonological working memory, but phoneme deletion remains unaffected. This suggests that different phonological awareness tasks do not equally rely on working memory. Nevo and Bar-Kochva (2015) found that simple phonological working memory tasks, such as digit repetition, real-word repetition, and pseudoword repetition, have a minimal impact on word recognition. However, complex phonological working memory tasks involving executive functioning produce effects. Nevertheless, these effects diminish by the age of five, indicating a decline in the impact of phonological working memory on word recognition with age. Cunningham et al. (2020) identified a dynamic relationship between phonological awareness, working memory, and reading. They categorized working memory tasks into those emphasizing serial recall and those requiring executive functioning, such as nonword repetition. Through cross-lag analysis, they observed that phonological awareness tasks are influenced by earlier phonological working memory, and phonological working memory indirectly impacts word recognition through the development of phonological awareness. Both types of earlier phonological working memory abilities directly influence word recognition development. However, serial recall, which requires less executive functioning, is not affected by earlier reading ability, whereas nonword repetition is influenced by earlier reading ability. This highlights how improving reading skills enhances phonological representation efficiency, thereby enhancing working memory, particularly for tasks requiring executive functioning. Given the above discussion, it is apparent that phonological working memory may affect word recognition, primarily through phonemic awareness, thereby indirectly influencing word recognition.

When discussing the correlation between phonological working memory and reading comprehension or vocabulary development, the relationship between Chinese working memory and these two aspects should be consistent with English (Leong et al., 2008). However, discussing the relationship between phonological working memory and Chinese character recognition should be less effective than in English (Liu et al., 2024; So & Siegel, 1997). This is because Chinese words' syllable length is generally shorter than in English, with most Chinese characters comprising a single syllable and morpheme. The demand for phonological working memory is lower in Chinese than in English.

1.13 Suggestions for Future Work

Intervention studies on word recognition for students with dyslexia should be the most urgent research topic. In English, phonological awareness, phonics, and fluency are evidence-based interventions (NICHHD, 2000). The effect sizes of different intervention methods are also well-documented (Hall et al., 2023). However, intervention studies for Chinese dyslexia are relatively scarce. Regarding phonological awareness intervention, significant improvements in character recognition ability have only been observed in Hong Kong, with very small effect sizes (Wang et al., 2017). Other interventions related to phonological awareness show the ability to enhance phonological awareness effectively but do not transfer to character recognition (Zhou et al., 2012). However, interventions focusing on phonetic radicals have shown positive effects (Ho & Ma, 1999; Ho et al., 2006; Shu et al., 2000), but as mentioned earlier, the reading abilities of participants during the intervention should be considered. Participants must reach certain reading abilities for phonetic radical instruction to be effective (Chen et al., 2016). Most morphological training also shows positive effects (Wu et al, 2009), but this should be considered about the participants' reading abilities.

In discussing the development of Chinese reading, various cognitive factors affecting reading ability should vary depending on reading proficiency. Past studies have mainly used chronological age as different stages, but a better model might differentiate based on a student's reading age (Hung et al., 2008; Wang et al., 2008). For example, regarding character identification, using the number of characters a student can recognize at different classification stages might provide a clearer understanding of the role of different cognitive factors. For instance, phonological awareness, orthographic rules, morphological awareness, and other cognitive factors—at which stage of ability do they exert their influence?

Additionally, reading ability can be distinguished into accuracy and fluency of character recognition, fluency, and reading comprehension. Future models will need to provide more detailed explanations of which cognitive factors influence different reading ability indicators at which stage of ability.

1.14 Conclusion

Although past research has shown many similarities between Chinese and English reading development, such as the function of phonetic radicals and their similarity to the regularity and consistency of English word recognition, this article emphasizes that for beginners, defined here as students below the third grade with knowing Chinese character fewer than 1200 words, they must first learn a certain number of characters or radicals before they can use phonetic radicals to obtain a word's pronunciation. Although compound characters are common for beginners, their phonetic functionality is poor. For students in Taiwan and mainland China, Zhuyin Fuhow and Hanyu Pinyin provide correct pronunciations, so learners do not need to derive pronunciation from the structure of characters. Because of these characteristics, the way cognitive factors operate and the timing of their effects must be considered.

References

Adams, M. J. (1994). *Beginning to read: Thinking and learning about print*. The MIT Press.

Ai, W. (1949). *Hanzi issues*. ZhongHua Book Company.

Carretti, B., Borella, E., Cornoldi, C., & De Beni, R. (2009). Role of working memory in explaining the performance of individuals with specific reading comprehension difficulties: A meta-analysis. *Learning and Individual Differences, 19*(2), 245–251. https://doi.org/10.1016/j.lindif.2008.10.002

Castles, A., & Coltheart, M. (1993). Varieties of developmental dyslexia. *Cognition, 47*(2), 149–180. https://doi.org/10.1016/0010-0277(93)90003-e

Chall, J. S. (1983). *Stages of reading development*. McGraw-Hill.

Chang, Y. J., Su, F. H., Tzeng, S. J., Ko, H. W., Wu, M. L., Yang, C. C., & Yang, C. Y. (2014). The contribution of rapid automatized naming to Chinese character recognition. *Learning and Individual Differences, 34*, 43–50. https://doi.org/10.1016/j.lindif.2014.05.010

Chen, H. C, Chang, L. Y. Chiou, Y. S., Sung, Y. T., & Chang, L. E. (2011). Chinese orthography database and its application in teaching Chinese characters. *Bulletin of Educational Psychology, 43*(Special Issue on Reading), 269–290

Chen, Y. C., Lian, Y. T., & Lee, J. R. (2016). Effectiveness of a remediation vocabulary program for second to third graders with low reading achievement. *NTTU Educational Research Journal, 27*(2), 51–76.

Cheung, H., Chen, H. C., Lai, C. Y., Wong, O. C., & Hills, M. (2001). The development of phonological awareness: Effects of spoken language experience and orthography. *Cognition, 81*(3), 227–241. https://doi.org/10.1016/S0010-0277(01)00136-6

Chung, K. K. H., & Ho, C. S. H. (2010). Dyslexia in Chinese language: An overview of research and practice. *Australian Journal of Learning Difficulties, 15*(2), 213–224. https://doi.org/10.1080/19404158.2010.495830

Coltheart, M. (1985). Cognitive neuropsychology and the study of reading. In M. I., Posner, & O. S. M. Marin, (Eds.), *Attention and performance XI* (pp. 3–37). Erlbaum.

Coltheart, M., Rastle, K., Perry, C., Langdon, R., & Ziegler, J. (2001). DRC: A dual route cascaded model of visual word recognition and reading aloud. *Psychological Review, 108*(1), 204–256. https://doi.org/10.1037/0033-295X.108.1.204

Cunningham, A. J., Burgess, A. P., Witton, C., Talcott, J. B., & Shapiro, L. R. (2020). Dynamic relationships between phonological memory and reading: A five year longitudinal study from age 4 to 9. *Developmental Science, 24*(1), Article e12986. https://doi.org/10.1111/desc.12986

DeFrancis, J. (1984). *The Chinese language: Fact and fantasy*. The University of Hawaii Press.

DeFrancis, J. (1989). *Visible speech: Diverse oneness of writing systems (Asian interactions and comparisons)*. University of Hawaii Press.

Dupuy, H. J. (1974). The rationale, development, and standardization of a basic word vocabulary test. Vital and Health Statistics. *Series 2. Data Evaluation and Methods Research, 60*, 1–85. https://doi.org/10.1037/e415662004-001

Frost R. (2012). Towards a universal model of reading. *Behavioral and Brain Sciences*, 1–67. https://doi.org/10.1017/S0140525X11001841

Gathercole, S. E., & Baddeley, A. D. (1993). Phonological working memory: A critical building block for reading development and vocabulary acquisition? *European Journal of Psychology of Education, 8*, 259–272. https://doi.org/10.1007/BF03174081

Hall, C., Dahl-Leonard, K., Cho, E., Solari, E. J., Capin, P., Conner, C. L., Henry, A. R., Cook, L., Hayes, L., Vargas, I., Richmond, C. L., & Kehoe, K. F. (2023). Forty years of reading intervention research for elementary students with or at risk for dyslexia: A systematic review and meta-analysis. *Reading Research Quarterly, 58*(2), 285–312. https://doi.org/10.1002/rrq.477

Ho, C. S., & Bryant, P. (1997a). Development of phonological awareness of Chinese children in Hong Kong. *Journal of Psycholinguistic Research, 26*(1), 109–126. https://doi.org/10.1023/a:1025016322316

Ho, C. S., & Bryant, P. (1997b). Phonological skills are important in learning to read Chinese. *Developmental Psychology, 33*(6), 946–951. https://doi.org/10.1037/0012-1649.33.6.946

Ho, C., Chan, D., Tsang, S.-M., Lee, S.-H., & Chung, K. (2006). Word learning deficit among Chinese dyslexic children. *Journal of Child Language., 33*, 145–161. https://doi.org/10.1017/S0305000905007154

Ho, C.S.-H., Chan, D.W.-O., Tsang, S.-M., & Lee, S.-H. (2002). The cognitive profile and multiple-deficit hypothesis in Chinese developmental dyslexia. *Developmental Psychology, 38*, 543–553. https://doi.org/10.1037/0012-1649.38.4.543

Ho, C. S., Chan, D. W., Lee, S. H., Tsang, S. M., & Luan, V. H. (2004). Cognitive profiling and preliminary subtyping in Chinese developmental dyslexia. *Cognition, 91*(1), 43–75. https://doi.org/10.1016/s0010-0277(03)00163-x

Ho, S. H. C., & Ma, N. L. (1999). Training in phonological strategies improves Chinese dyslexic children's character reading skills. *Journal of Research in Reading, 22*(2), 131–142. https://doi.org/10.1111/1467-9817.00078

Ho, C. S.-H., Ng, T.-T., & Ng, W.-K. (2003). A "radical" approach to reading Development in Chinese: The role of semantic radicals and phonetic radicals. *Journal of Literacy Research, 35*(3), 849–878. https://doi.org/10.1207/s15548430jlr3503_3

Hsiang, T.P., Graham, S., Wang, Z., & Gong, Y., (2021). How Chinese characters are taught: An analysis of three popular textbooks used in Macao. *Technology Knowledge Learning, 26*, 355–381 (2021). https://doi.org/10.1007/s10758-021-09492-8

Huang, H. S., & Hanley, J. R. (1995). Phonological awareness and visual skills in learning to read Chinese and English. *Cognition, 54*(1), 73–98. https://doi.org/10.1016/0010-0277(94)00641-W

Hung, L. Y., Wang, C. C., Chang, Y. W., & Chen, H. F. (2008). Development of assessment of Chinese character lists for graders. *Psychological Testing, 55*(3), 489–508. https://www.airitilibrary.com/Common/Click_DOI?DOI=10.7108%2fPT.200812.0003

Huang, F. S. (1994). *Research on illiterate citizens' literacy standards and literacy vocabulary in our country*. National Taiwan Normal University.

Liberman, I. Y., Shankweiler, D., Fischer, F. W., & Carter, B. (1974). Explicit syllable and phoneme segmentation in the young child. *Journal of Experimental Child Psychology, 18*, 201–212. https://doi.org/10.1016/0022-0965(74)90101-5

Landerl, K., Wimmer, H., & Frith, U. (1997). The impact of orthographic consistency on dyslexia: A German-English comparison. *Cognition, 63*(3), 315–334. https://doi.org/10.1016/s0010-0277(97)00005-x

Lee, C. Y., Tsai, J. L., Su, C. I., Tzeng, O. J. L., & Hung, D. L. (2005). Consistency, regularity, and frequency effects in naming Chinese characters. *Language and Linguistics, 7*(3), 619–635.

Lee, J. R., Kuo, W. J., Lee, R. W., & Tzeng, O. J. L. (2024). Visualizing cross-linguistic reading brains: Universality and interactive specificities. *Journal of Chinese Linguistics, 52*(1), 1–28. https://doi.org/10.1353/jcl.2024.a919397

Li, L., Wang, H. C., Castles, A., Hsieh, M. L., & Marinus, E. (2018). Phonetic radicals, not phonological coding systems, support orthographic learning via self-teaching in Chinese. *Cognition, 176*, 184–194. https://doi.org/10.1016/j.cognition.2018.02.025

Li, L., Marinus, E., Castles, A., Hsieh, M.-L., & Wang, H.-C. (2021). Semantic and phonological decoding in children's orthographic learning in Chinese. *Scientific Studies of Reading, 25*(4), 319–334. https://doi.org/10.1080/10888438.2020.1781863

Liu, S., Li, N., Zhang, X., & Wang, L. C. (2024). Effects of working memory and visual search skill on Chinese reading comprehension: Examining the simple view of reading. *Read and Writing*. https://doi.org/10.1007/s11145-024-10515-w

Liu, P. D., & McBride-Chang, C. (2010). What is morphological awareness? Tapping lexical compounding awareness in Chinese third graders. *Journal of Educational Psychology, 102*(1), 62–73. https://doi.org/10.1037/a0016933

Logan, J. A., Schatschneider, C., & Wagner, R. K. (2011). Rapid serial naming and reading ability: The role of lexical access. *Reading and Writing, 24*(1), 1–25. https://doi.org/10.1007/s11145-009-9199-1

Lonigan, C. J., Burgess, S. R., Anthony, J. L., & Barker, T. A. (1998). Development of phonological sensitivity in 2- to 5-year-old children. *Journal of Educational Psychology, 90*(2), 294–311. https://doi.org/10.1037/0022-0663.90.2.294

Leong, C. K., Tse, S. K., Loh, K. Y., & Hau, K. T. (2008). Text comprehension in Chinese children: Relative contribution of verbal working memory, pseudoword reading, rapid automatized naming, and onset-rime phonological segmentation. *Journal of Educational Psychology, 100*(1), 135–149. https://doi.org/10.1037/0022-0663.100.1.135

Ma, W. (2014). Corpus-based analysis of semantic transparency between high frequent English and Chinese compounds. *Theory and Practice in Language Studies, 4*(6), 1138–1143. https://doi.org/10.4304/tpls.4.6.1138-1143

McBride-Chang, C. A. (2016). Is Chinese special? Four aspects of Chinese literacy acquisition that might distinguish learning Chinese from learning alphabetic orthographies. *Educational Psychology Review, 28*(4), 523–549. https://doi.org/ https://doi.org/10.1007/s10648-015-9318-2

McBride-Chang, C., Bialystok, E., Chong, K. K. Y., & Li, Y. (2004). Levels of phonological awareness in three cultures. *Journal of Experimental Child Psychology, 89*, 93–111. https://doi.org/10.1016/j.jecp.2004.05.001

McBride-Chang, C., Cho, J. R., Liu, H., Wagner, R. K., Shu, H., Zhou, A., Cheuk, C. S., & Muse, A. (2005). Changing models across cultures: Associations of phonological awareness and morphological structure awareness with vocabulary and word recognition in second graders from Beijing, Hong Kong, Korea, and the United States. *Journal of Experimental Child Psychology, 92*(2), 140–160. https://doi.org/10.1016/j.jecp.2005.03.009

McBride-Chang, C., & Kail, R. V. (2001). Cross-cultural similarities in the predictors of reading acquisition. *Child Development, 73*(5), 1392–1407. https://doi.org/10.1111/1467-8624.00479

McBride-Chang, C., Shu, H., Zhou, A., Wat, C. P., & Wagner, R. K. (2003). Morphological awareness uniquely predicts young children's Chinese character recognition. *Journal of Educational Psychology, 95*(4), 743–751. https://doi.org/10.1037/0022-0663.95.4.743

McBride, C., & Wang, Y. (2015). Learning to read Chinese: Universal and unique cognitive cores. *Child Development Perspectives, 9*(3), 196–200. https://doi.org/10.1111/cdep.12132

McBride, C., Wang, Y., & Cheang, L.M.-L. (2018). Dyslexia in Chinese. *Current Developmental Disorders Reports, 5*(4), 217–225. https://doi.org/10.1007/s40474-018-0149-y

Melby-Lervåg, M., Lyster, S. A., & Hulme, C. (2012). Phonological skills and their role in learning to read: A meta-analytic review. *Psychological Bulletin, 138*(2), 322–352. https://doi.org/10.1037/a0026744

Morais, J., Bertelson, P., Cary, L., & Alegria, J. (1986). Literacy training and speech segmentation. *Cognition, 24*, 45–64.

Nevo, E., & Bar-Kochva, I. (2015). The relations between early working memory abilities and later developing reading skills: A longitudinal study from kindergarten to fifth grade. *Mind, Brain, and Education, 9*(3), 154–163. https://doi.org/10.1111/mbe.12084

National Institute of Child Health and Human Development. (2000). *Teaching children to read: An evidence-based assessment of the scientific research literature on reading and Its implications for reading instruction*. National Institute of Child Health and Human Development.

Oakhill, J. V., & Cain, K. (2012). The precursors of reading ability in young readers: Evidence from a four-year longitudinal study. *Scientific Studies of Reading, 16*(2), 91–121. https://doi.org/10.1080/10888438.2010.529219

Oakhill, J., & Kyle, F. (2000). The relation between phonological awareness and working memory. *Journal of Experimental Child Psychology, 75*(2), 152–164. https://doi.org/10.1006/jecp.1999.2529

Pan, D. J., Meng, X., Lee, J. R., Ng, C. Y., & McBride, C. (2024). The cognitive-linguistic profiles and academic performances of Chinese children with dyslexia across cultures: Beijing, Hong Kong, and Taipei. *Annual of Dyslexia.* https://doi.org/10.1007/s11881-024-00304-z

Paulesu, E., Démonet, J. F., Fazio, F., McCrory, E., Chanoine, V., Brunswick, N., Cappa, S. F., Cossu, G., Habib, M., Frith, C. D., & Frith, U. (2001). Dyslexia: cultural diversity and biological unity. *Science, 291*(5511), 2165–2167. https://doi.org/10.1126/science.1057179

Peng, P., Wang, C., Tao, S., & Sun, C. (2017). The deficit profiles of Chinese children with reading difficulties: A meta-analysis. *Educational Psychology Review, 29*(3), 513–564. https://doi.org/10.1007/s10648-016-9366-2

Perfetti, C., Cao, F., & Booth, J. (2013). Specialization and universals in the development of reading skill: How Chinese research informs a universal science of Reading. *Scientific Studies of Reading, 17*(1), 5–21. https://doi.org/10.1080/10888438.2012.689786

Reichle, E. D., & Yu, L. (2018). Models of Chinese reading: Review and analysis. *Cognitive Science, 42*(Suppl 4), 1154–1165. https://doi.org/10.1111/cogs.12564

Ruan, Y., Georgiou, G. K., Song, S., Li, Y., & Shu, H. (2018). Does writing system influence the associations between phonological awareness, morphological awareness, and reading? *A Meta-Analysis. Journal of Educational Psychology, 110*(2), 180–202. https://doi.org/10.1037/edu0000216

Shu, H., & Anderson, R. C. (1997). Role of radical awareness in the character and word acquisition of Chinese children. *Reading Research Quarterly, 32*(1), 78–89. https://doi.org/10.1598/RRQ.32.1.5

Shu, H., Anderson, R. C., & Wu, N. (2000). Phonetic awareness: Knowledge of orthography–phonology relationships in the character acquisition of Chinese children. *Journal of Educational Psychology, 92*(1), 56–62. https://doi.org/10.1037/0022-0663.92.1.56

Shu, H., Chen, X., Anderson, R. C., Wu, N., & Xuan, Y. (2003). Properties of school Chinese: Implications for learning to read. *Child Development, 74*(1), 27–47. https://doi.org/10.1111/1467-8624.00519

Shu, H., McBride-Chang, C., Wu, S., & Liu, H. (2006). Understanding Chinese developmental dyslexia: Morphological awareness as a core cognitive construct. *Journal of Educational Psychology, 98*(1), 122–133. https://doi.org/10.1037/0022-0663.98.1.122

Shu, H., Peng, H., & McBride-Chang, C. (2008). Phonological awareness in young Chinese children. *Developmental Science, 11*(1), 171–181. https://doi.org/10.1111/j.1467-7687.2007.00654.x

Siok, W. T., & Fletcher, P. (2001). The role of phonological awareness and visual-orthographic skills in Chinese reading acquisition. *Developmental Psychology, 37*(6), 886–899. https://doi.org/10.1037/0012-1649.37.6.886

Snow, C. E., Burns, M. S., & Griffin, P. (Eds.). (1998). *Preventing reading difficulties in young children.* National Academy Press.

Song, S., Georgiou, G. K., Su, M., & Hua, S. (2016). How well do phonological awareness and rapid automatized naming correlate with Chinese reading accuracy and fluency? A meta-analysis. *Scientific Studies of Reading, 20*, 99–123. https://doi.org/10.1080/10888438.2015.1088543

Stanovich, K. E., Cunningham, A. E., & Cramer, B. B. (1984). Assessing phonological awareness in kindergarten children: Issues of task comparability. *Journal of Experimental Child Psychology, 38*(2), 175–190. https://doi.org/10.1016/0022-0965(84)90120-6

So, D., & Siegel, L. S. (1997). Learning to read Chinese: Semantic, syntactic, phonological and working memory skills in normally achieving and poor Chinese readers. *Reading and Writing: An Interdisciplinary Journal, 9*(1), 1–21. https://doi.org/10.1023/A:1007963513853

Tan, L. H., Spinks, J. A., Eden, G. F., Perfetti, C. A., & Siok, W. T. (2005). Reading depends on writing, in Chinese. *Proceedings of the National Academy of Sciences of the United States of America, 102*(24), 8781–8785. https://doi.org/10.1073/pnas.0503523102

Tong, X., & McBride-Chang, C. (2010). Developmental models of learning to read Chinese words. *Developmental Psychology, 46*(6), 1662–1676. https://doi.org/10.1037/a0020611

Tzeng, S. J., Jian, S. J., Chang, Y. T., Chou, L. L., & Lian, Y. L. (2005). The prediction of Chinese reading and word recognition by using early naming speed and phonological awareness: A four-year longitudinal study. *Bulletin of Special Education, 28*, 123–144.

Tzeng, Y. L., & Lee, C. Y. (2012). The developmental trajectories of orthography-to-phonology mapping consistency in learning to read Chinese. *Contemporary Educational Research Quarterly, 20*(4), 45–84. https://doi.org/10.6151/CERQ.2012.2004.02

Tseng, C. C., Hu, J. F., Chang, L. Y., & Chen, H. C. (2022). Learning to read Chinese: The roles of phonological awareness, paired-associate learning, and phonetic radical awareness. *Reading and Writing, 36*, 1769–1795. https://doi.org/10.1007/s11145-022-10352-9

Wang, C. C., Hung, L. Y., Chang, Y., & Chen, H. (2008). The number of characters school students know from grade 1 to 9. *Bulletin of Education Psychology, 39*(4), 555–568.https://doi.org/10.6251/BEP.20071026

Wang, Li. (1996). *History of Chinese Linguistics*. Wu-Nan Book Inc.

Wang, L. C., Liu, D., Chung, K. K. H., & Yang, H. M. (2017). Development of lexical tone awareness in Chinese children with and without dyslexia. *Contemporary Educational Psychology, 49*, 203–214. https://doi.org/10.1016/j.cedpsych.2017.02.002

Wu, X., Anderson, R. C., Li, W., Wu, X., Li, H., Zhang, J., Zheng, Q., Zhu, J., Shu, H., Jiang, W., Chen, X., Wang, Q., Yin, L., He, Y., Packard, J., & Gaffney, J. S. (2009). Morphological awareness and Chinese Children's literacy development: An intervention study. *Scientific Studies of Reading, 13*(1), 26–52. https://doi.org/10.1080/10888430802631734

Wu, J., Tsai, Y., & Lin, C. (2018). *Survey report of commonly used words and phrases from 2011 to 2015*. National Academy for Educational Research, Taipei, Taiwan.

Xu, S. (1965). *ShuoWenJieZi DuanZhu (Duan's annotation of Shuo Wen Jie Zi)*. YeeWen Publishing Company.

Yang, L.-Y., Guo, J.-P., Richman, L. C., Schmidt, F. L., Gerken, K. C., & Ding, Y. (2013). Visual skills and Chinese reading acquisition: A meta-analysis of correlation evidence. *Educational Psychology Review, 25*(1), 115–143. https://doi.org/10.1007/s10648-013-9217-3

Yopp, H. K. (1988). The validity and reliability of phonemic awareness tests. *Reading Research Quarterly, 23*(2), 159–177. https://doi.org/10.2307/747800

Zhang, L., Xia, Z., Zhao, Y., Shu, H., & Zhang, Y. (2023). Recent advances in Chinese developmental dyslexia. *Annual Review of Linguistics, 9*(1), 439–461. https://doi.org/10.1146/annurev-linguistics-030421-065648

Zhou, Y.-L., McBride-Chang, C., Fong, C.Y.-C., Wong, T.T.-Y., & Cheung, S. K. (2012). A comparison of phonological awareness, lexical compounding, and homophone training for Chinese word reading in Hong Kong kindergartners. *Early Education and Development, 23*(4), 475–492. https://doi.org/10.1080/10409289.2010.530478

Zhou, L. & Perfetti, C. (2021). Consistency and regularity effects in character identification: A greater role for global than local mapping congruence. *Brain and Language, 221*, 104997. https://doi.org/10.1016/j.bandl.2021.104997

Zhou, L., & Perfetti, C. (2023). Global mapping congruence influences phonological interference: ERP evidence for early and persistent ortho-phonological effects in reading Chinese for meaning. *Journal of Experimental Psychology: Learning, Memory, and Cognition, 49*(7), 1176–1191. https://doi.org/10.1037/xlm0001221

Zhou, Y.-G. (1978). To what degree are the "phonetics" of present-day Chinese characters still phonetic? *Zhongguo Yuwen, 146*, 172–177.

Chapter 2
Neural Signatures for the Statistical Learning of Chinese Language Processing

Chia-Ying Lee, Pei-Chun Chao, and Yu-Lin Tzeng

Abstract Humans are known to continuously extract regularities from the flow of visual and auditory stimulation to perceive the world's structure, decrease uncertainty, and reduce information load. The Bimodal interactive activation model (BIAM) assumes that the statistical learning mechanism is the core ability to extract orthographic regularities to map onto the phonological (i.e., phoneme and rime) units within words. As a consequence, literacy may also shape speech processing. Studies of alphabetic writing systems have demonstrated the effects of bidirectional mapping consistency between orthography and phonology in both visual and spoken word recognition. However, writing systems vary in orthographic depth, which may result in developmental differences. Chinese is characterized as an ideographic writing system with deep orthography. This chapter will review a series of behavioral and neuroimaging studies to investigate how children acquire Chinese orthographic knowledge, such as radical position, phonetic regularity, and consistency, in learning to read Chinese. Meanwhile, the neural evidence for the homophone density and orthographic consistency effects in spoken word recognition also suggests that literacy experience fundamentally changes the preexisting phonological representation in the mental lexicon. This cross-linguistic evidence supports that, although languages naturally differ in their scripts and orthographic principles, a reading model can be universal to reflect the common cognitive operations across writing systems.

Keywords Chinese reading acquisition · Spoken word recognition · Cross-code consistency · Statistical learning · Event-related potentials

C.-Y. Lee (✉) · P.-C. Chao · Y.-L. Tzeng
Brain and Language Laboratory, Academia Sinica, Taipei, Taiwan
e-mail: chiaying@gate.sinica.edu.tw

Institute of Linguistics, Academia Sinica, Taipei, Taiwan

D. H. Wu (ed.), *Cognitive and Neural Foundations of Chinese Reading*, Chinese Language Learning Sciences, https://doi.org/10.1007/978-981-96-6669-0_2

2.1 Introduction

Understanding the interconnections between spellings (orthography), sounds (phonology), and meanings (semantics) is crucial for a well-established word recognition model. Speech predates written language throughout human history and individual development (Liberman, 1992). Learning spoken language mainly entails building the connection between phonology and semantics. For children to achieve literacy, they need to learn the specific orthographic codes of a written language and try to figure out the rules of how these codes correspond to spoken words and meanings. Therefore, it is natural to expect phonology to have a pivotal influence on learning to read. On the contrary, questioning the activation of words' written forms during listening seems illogical, as it offers no apparent benefit in comprehending spoken language. Nevertheless, there is increasing evidence showing that literacy shapes speech processing.

However, writing systems vary in the transparency of spelling-to-sound consistency. In alphabetic writing systems, there is a spectrum ranging from shallow orthographies (e.g., Finnish and Welsh), where each letter directly maps onto one phoneme, and conversely, to deep orthographies (e.g., English), where a single letter or group of letters can represent different pronunciations and a phoneme may be spelled in multiple ways. Ziegler and Goswami (2005) introduced the psycholinguistic grain size theory to explain variations in how reading skills develop among different languages. Learning to read is finding the optimal mapping and the shared grain size between phonology and orthography. Thus, orthographic transparency may affect the development of grapheme-phoneme recoding skills, leading to diversity in the evolution of reading approaches. Specifically, children should learn to read more quickly in writing systems with highly consistent letter-sound correspondences. In contrast to English, where words consist of letters representing phonemes, Chinese characters, as a basic writing unit, are composed of radicals and map onto one syllable rather than a phoneme. Due to the absence of phonemic representation in Chinese orthography, one cannot phonetically assemble the pronunciation of a character based on its orthographic components. Therefore, finding the optimal mapping grain size has been the major challenge for learning to read Chinese.

Increasing evidence has indicated a universal cognitive framework for reading and its acquisition across languages, although differences in orthographic transparency may influence specific details. (Frost, 2012; Perfetti & Harris, 2013; Perfetti et al., 2013; Ziegler & Goswami, 2005). However, most evidence stemmed from studies on the alphabetic writing system. Therefore, cross-linguistic research has used the highly contrasting Chinese writing system to explore the universal features of word recognition models. This chapter aims to comprehensively review behavioral and neural evidence for the bidirectional mappings between orthography and phonology in Chinese word recognition.

2.2 The Bimodal Interactive Activation Model (BIAM) for Visual and Auditory Word Recognition

Word recognition research has shown that identifying a word is influenced by the characteristics of its neighbors, such as orthographic neighborhood size (i.e., the number of words that can be created by changing a single letter in a target word; Andrews, 1989; Grainger et al., 1989), orthography-to-phonology consistency (i.e., the degree to which similarly spelled words are pronounced similarly; Taraban & McClelland, 1987), and phonological neighborhood density (i.e., the number of words that can be derived by changing one phoneme in a target word; Vitevitch et al., 1999). Grainger and Ziegler (2008) proposed the bimodal interactive activation model (BIAM) to account for these effects. This model proposes three essential codes in the architecture similar to the triangle model (Seidenberg & McClelland, 1989): orthography, phonology, and semantics, which proposes that word recognition involves an ongoing and dynamic learning process, contingent upon the interplay among the new inputs, the present system state, and the experience. The BIAM adds a lexical and sublexical distinction to the framework to separate the orthographic and phonological codes into two processing pools (see Fig. 2.1). One represents entire words, and the other breaks words down into smaller orthographic and phonological units, such as single or multi-letter combinations, phonemes, or combinations of phonemes. Additionally, BIAM posits that visual and spoken word recognition builds on interactively activating orthographic, phonological, and semantic codes in both directions. Cross-code consistency indicates the degree of compatibility or coherence among all concurrently activated representations across orthography, phonology, and semantics, which is essential for stable word recognition (Grainger & Ziegler, 2008).

According to BIAM, how consistently orthography (O) matches phonology (P) can be assessed in both directions. For visual word recognition, the mapping consistencies from O to P and P to O have been termed *feedforward* and *feedback consistency* (see Fig. 2.2). The feedforward consistency (O to P) effects have been observed in numerous studies. People were faster and more accurate at naming highly feedforward-consistent words (e.g., words ending in -ean like the bean, lean, and dean are always pronounced with /in/). (Jared, 1997, 2002; Taraban & McClelland, 1987).

The feedback consistency (or the so-called orthographic consistency) refers to whether words have onsets or rhymes that can be spelled in multiple ways. For example, in English, the rime/ip/ can be spelled either -eep or -eap, as in reap and weep. Conversely, the rime /ʌk/ is always spelled with the letters -uck, as in duck and truck (see Fig. 2.3). Stone et al. (1997) observed a significant increase in visual lexical decision time for words with feedback inconsistency compared to those with feedback consistency. Besides, recognition errors were higher for feedback-inconsistent words (9.8%) than for feedback-consistent words (3.3%). Ziegler et al. (1997) also found the feedback consistency effect in French, despite its difficulty deriving pronunciation from spelling, in lexical decision and naming tasks. These findings revealed the feedback influence from phonology to orthography on word recognition, challenging

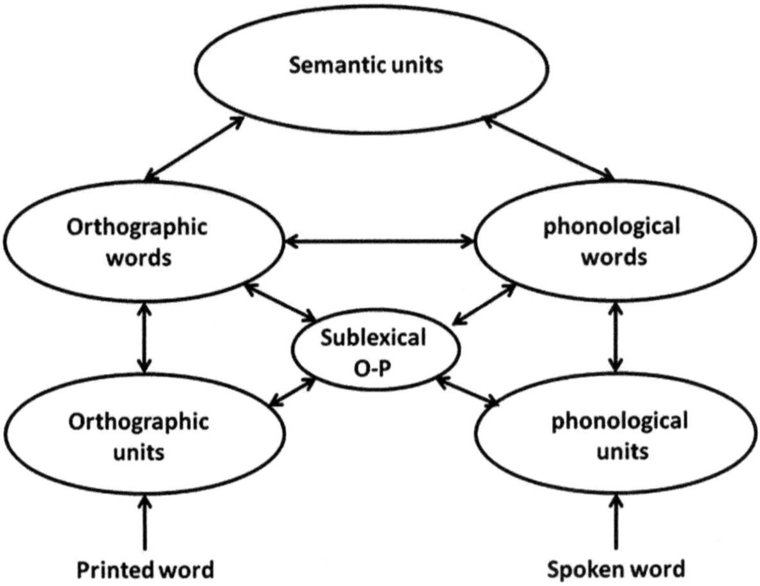

Fig. 2.1 The bimodal interactive activation model (BIAM) architecture for word recognition and the schema for bidirectional activation between orthography and phonology

Fig. 2.2 Feedforward and feedback consistency for visual word recognition

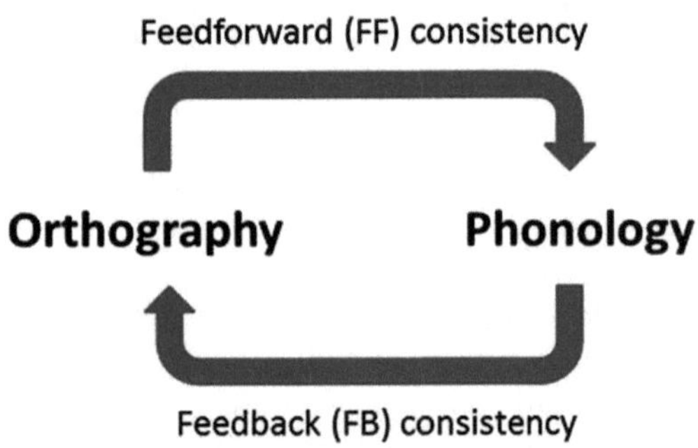

traditional theories that recognize words in a one-way bottom-up process. (Norris et al., 2000).

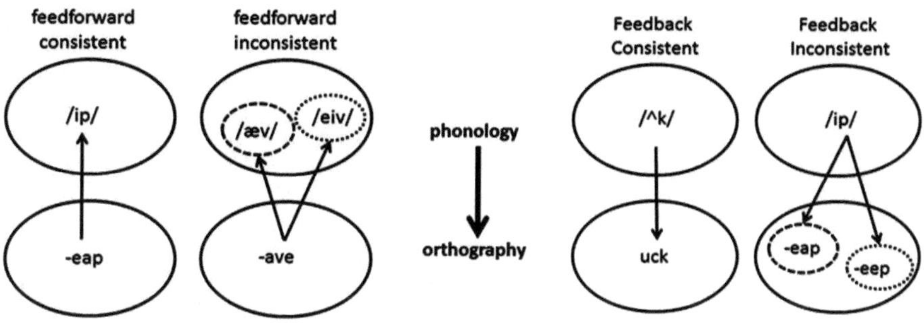

Fig. 2.3 Examples of feedforward and feedback consistency in English

2.3 The Cross-Code Consistency Framework for Chinese: Feedforward Consistency

Chinese is often classified as a logographic writing system. Chinese characters are composed of radicals and map onto a single-syllable morpheme rather than a phoneme in spoken language. Chang et al. (2016) analyzed traditional Chinese character databases. They revealed that there are more than 80% of Chinese characters are phonograms (i.e., 踩, /cai3/, "to step on"), which consist of a semantic radical (i.e., 足, /zu2/, "foot") and a phonetic radical (i.e., 采, /cai3/, "gathering") to provide clues for character's meaning and pronunciation. Meanwhile, around 72% of phonograms have a left–right structure. Around 90% of the left–right structured phonograms are composed of the semantic radicals(S) on the left and phonetic radicals(P) on the right (i.e., 踩, an SP character). Only 10% have phonetic radicals on the left side (i.e., 彩, a PS character). To examine the role of the dominant distribution of SP characters in learning to read Chinese phonograms, Lo et al. (2007) test how children name Chinese pseudocharacters with SP or PS structures. They found that beginning readers tended to name the pronounceable radical regardless of its position. As their reading ability increased (measured by a Chinese character recognition test), the children tended to use the radical on the right side to infer the pronunciation. The findings support that Chinese readers gradually acquire the metalinguistic knowledge of Chinese orthography, such as the statistical properties of a radical's position and the conveyance of semantic and phonological information by radicals as they learn to read Chinese.

In Chinese, the phonological relationship between a phonogram and its phonetic radical can be addressed by ***regularity*** or ***consistency***. Regularity refers to whether a character and its phonetic radical share the same pronunciation. For example, 楓 /feng1/ is a regular character because it is pronounced the same as its phonetic radical 風 /feng1/, whereas 猜 /tsai1/ is an irregular phonogram as it is pronounced differently from its phonetic radical 青/qing1/. Previous studies have demonstrated the frequency-regularity interaction in naming Chinese phonograms. The reaction time for naming a regular character is much faster than for naming an irregular character, especially for low-frequency characters (Hue, 1992; Lee et al., 2005;

Fig. 2.4 Examples of the Feedforward consistency in Chinese

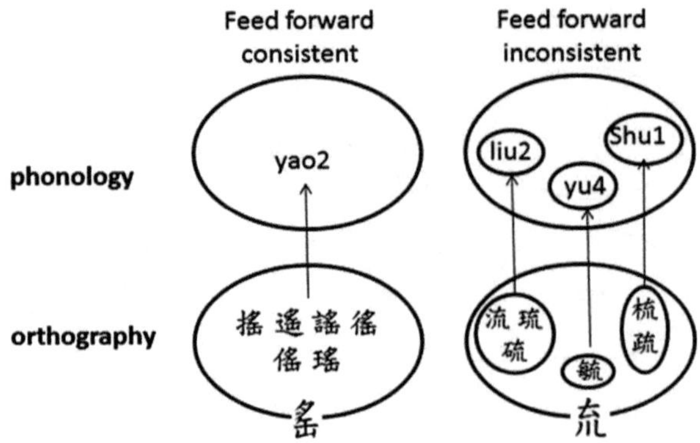

Seidenberg, 1985). Those data suggest that Chinese phonograms are not read via a direct association between orthography and phonology, but they involve sublexical processes.

However, a large set of phonograms carries phonetic radicals that are no longer legitimate characters by themselves. Studies alternatively applied the **feedforward consistency** to measure statistical relationships between orthography and phonology. The O-to-P consistency of Chinese phonograms is measured as the degree of reliability of a phonetic radical to provide a phonological clue or whether the pronunciation of a character agrees with those of its orthographic neighbors containing the same phonetic radical (see Fig. 2.4). For example, 搖 /yao2/ has six orthographic neighbors, namely, 鷂, 瑤, 遙, 徭, 傜, and 謠. All the neighbors are "friends" of 搖 because they share the same pronunciation. On the other hand, 流 /liu2/ also has five orthographic neighbors; only 琉 /liu2/ and 硫 /liu2/ are its "friends," but not 梳 /shu1/, 疏 /shu1/ and 毓 /yu4/. Therefore, 搖 and its orthographic neighbors are high-consistency characters (consistency index = 1), whereas 流, 琉, and 硫 are low-consistency characters (consistency index = 0.33).

The definition of consistency in English and Chinese is parallel concerning the representation of the statistical relationship between orthographic forms and their pronunciations. Studies used in the naming task have also demonstrated the consistency effect in naming Chinese, in which high-consistent characters were named faster with higher accuracy than low-consistent characters (Fang et al., 1986; Hue, 1992; Yang et al., 2009). Lee et al. (2005) manipulated the feedforward consistency measure by token (i.e., considering character frequencies) in a naming study and demonstrated consistency effects for both high and low-frequency characters and a significant interaction between consistency and frequency. Congruent with previous studies, the consistency effect was more prominent in low-frequency than in high-frequency words. The consistency effect has also been found in homophone judgment (Chiu et al., 2016; Lee et al., 2006, 2007) and in reading sentences (Tsai et al., 2004), even in Taiwanese deaf signers (Chiu et al., 2016). The corpus-based analysis also demonstrated how phonetic consistency interplays with other lexical properties in Chinese character recognition. For example, Chang et al. (2016) established a large-scale psycholinguistic norm of 3,314 traditional Chinese characters with their lexical properties and naming responses collected from 140 Chinese native speakers. The multiple-regression analysis showed that the naming latencies were affected by frequency, consistency, and regularity while controlling the effects from a set of lexical and semantic variables, which included frequency, regularity, familiarity, consistency, number of strokes, homophone density, semantic ambiguity rating, phonetic combinability, semantic combinability, and the number of disyllabic compound words formed by a character. These findings suggest that Chinese readers are sensitive to the statistical mapping consistency between Chinese orthography and phonology.

Tzeng and Lee (2012) provided developmental evidence for statistical learning in learning to read Chinese. This study invited Children from 4th Grade to 6th Grade to name a set of phonograms, which were divided into two consistency levels (high and low) and three types of phonograms (regular character, irregular character, and independent phonograms). The data revealed that the phonetic consistency effect emerged in 4th Grade when reading regular phonograms. The consistency effect was not found in reading irregular phonograms until 5th Grade. By 6th Grade, the consistency effect was observed in reading all three types of phonograms. These findings imply that the feedforward consistency effect in reading Chinese is an emergent property of vocabulary growth and supports the statistical learning perspective for reading acquisition (Lee, 2009).

Further evidence comes from studies using EEG and MEG techniques to trace the time courses of utilizing various lexical and sublexical properties for character reading (Hsu et al., 2009; Lee et al., 2006, 2007). Lee et al. (2007) investigated the temporal dynamics of the consistency effect in reading Chinese phonograms by asking participants to do a homophone judgment task. The result demonstrated that the consistency effect influenced N170, P200, and N400. Low consistent characters elicited greater N170 amplitudes in the temporal-occipital region, greater P200 amplitudes in the frontal region, and less negativity on N400 in the central region than did high consistent characters. These findings suggest that low-consistency characters produce a greater activation for the initial analysis of the orthographical and phonological representations (indexed by P200), while high-consistency characters involve a greater lexical competition in the later stage (indexed by N400; Lee et al., 2007).

Hsu et al. (2009) further examined the statistical learning perspective for learning to read Chinese by manipulating phonetic consistency and phonetic combinability. The phonetic combinability is defined as the number of characters sharing the same phonetic radical, namely the orthographic neighborhood. According to the statistical learning hypothesis, the representation of phonetic consistency should implicitly emerge. Therefore, a Chinese character with a larger orthographic neighborhood may obtain more experience detecting the mapping principle between orthography and phonology. In other words, phonetic consistency should be more salient when reading high-combinability characters. Indeed, this study demonstrated the consistency-by-combinability interaction on N170, P200, and N400. The phonetic consistency effect is mainly found in phonograms with large phonetic combinability (Hsu et al., 2009). Hsu et al. (2014) applied source analysis techniques with MEG recording. They demonstrated a combinability effect in the right fusiform gyrus at ~170 ms, which may reflect perceptual expertise in processing Chinese orthography. At ~200 ms, the left anterior insular cortex showed larger activity in reading small combinability characters than in reading large combinability characters, and the left inferior parietal cortex showed larger activity in reading low-consistency characters than in reading high-consistency characters. These results indicate that the left anterior insula cortex and inferior parietal cortex may participate in the early stages of feature analysis and integrating orthographic and phonological information in reading Chinese phonograms.

Lee and colleagues also conducted a series of fMRI studies by manipulating the character frequency, phonetic consistency, and lexicality to reveal the neural correlates for Chinese orthography-to-phonology transformation. These studies identified a set of neural correlates, including the left inferior frontal gyrus, the left temporoparietal (inferior parietal gyrus and supramarginal gyrus) region, and the left temporo-occipital junction, involved in Chinese character reading (Lee et al., 2004, 2010). These regions are similar to the neural substrates responsible for reading alphabetic writing systems (Bolger et al., 2008; Weiss et al., 2018). Although Chinese and alphabetic languages differ in orthographic transparency, evidence from imaging studies supports that the neural mechanism of mapping consistency between orthography and phonology is universal across languages.

2.4 How Literacy Shapes Chinese Spoken Language Processing: Feedback Consistency in Chinese

Acquiring literacy is learning how orthographic units connect with previously established phonological and semantic representations. The automatic phonological activation in reading is naturally expected and well-demonstrated in visual word recognition. Thus, there seems to be no apparent advantage in engaging a word's orthography while recognizing spoken words because the connection between phonology and semantics is already formed before becoming literate. However, BIAM suggests that written forms, sounds, and meanings all activate and influence each other in a bidirectional way, regardless of whether the word is seen or heard. The literacy experience may alter how a person recognizes spoken words. Recent evidence supports that the development of orthographic knowledge plays an essential role in speech processing. For example, when illiterates struggle to delete or add the initial phoneme of nonwords, literates have become aware of each constitutive phonemic entity in syllables (Morais et al., 1979). Subsequently, Ehri and Wilce (1980) demonstrated that skilled children, when presented with the words PITCH and RICH, would identify an additional phoneme in /pɪtʃ/ compared to /rɪtʃ/ due to the varying number of letters. Meanwhile, Seidenberg and Tanenhaus (1979) observed that making a rhyme judgment for two spoken words is more difficult when their spellings are distinct (e.g., rye–pie) than when they share the same spelling (e.g., pie–tie), the so-called orthographic consistency effect. These observations suggested that the reading experience reshapes spoken language processing.

Studies used ERP to examine when and how orthographic consistency affects spoken word recognition. Namely, if the P-to-O consistency effect is observed at the N400, it implies that orthography affects the access to lexical retrieval. In contrast, orthographic information disturbs the post-lexical processing, including decision-making or memory retrieval, if the P-to-O consistency effect occurs at the late positive component (LPC). For instance, Perre and Ziegler (2008) manipulated the P-to-O consistency either at an early or a late position of spoken words and found the P-to-O consistency effect on the N400 in an auditory lexical decision task. Compared to the consistent words, the inconsistent words produced larger negativities with central-posterior distribution between 300 and 350 ms. The P-to-O consistency effects were locked to around 200 ms after the onset of inconsistency in spoken words. Pattamadilok et al. (2009) also showed the P-to-O consistency effects on N400s in a semantic go/no-go task. Notably, the P-to-O consistency effect occurred ahead of the word frequency effect, which served as a marker for lexical entry. These ERP findings revealed that orthographic knowledge automatically influences the fundamental lexical processes during spoken word recognition.

Findings from neuroimaging studies also provide insights into how literacy acquisition alters the processing of phonological information in the brain. More precisely, the P-to-O consistency effect is expected to emerge within the speech network involving the insula, left inferior frontal gyrus (IFG), supramarginal gyrus (SMG), and superior temporal gyrus (STG) if learning to read modifies the organization and the qualitative development of preexisting phonological representation. For instance, Montant et al. (2011) observed that lexical decisions to speak words with low P-to-O consistency evoked greater fMRI activation in the left insula and IFG than those with high P-to-O consistency. Alternatively, if learning to read would establish robust and enduring links between preexisting phonological and newly acquired orthographic representations, the P-to-O consistency effect would be found in the left ventral occipitotemporal cortex (vOT). Other fMRI studies demonstrated increased activation in the left fusiform during auditory spelling or rhyming judgment tasks compared to perceptual tone tasks (Booth et al., 2002, 2007; Yoncheva et al., 2010).

While there is abundant evidence regarding the consistency of O-to-P mapping in Chinese reading, only a limited number of studies have delved into the impact of P-to-O mapping consistency on Chinese spoken visual word recognition. In Chinese, around 1100 syllables are mapped onto around 5,000 different characters. According to the Academia Sinica balanced corpus (Huang & Chen, 1998), given the elementary consonant–vowel structure for most syllables, approximately 1300 monosyllables are allocated to around 5000 various traditional Mandarin characters. The prevalence of homophony in Chinese suggests that orthography could exert a more pronounced effect on spoken word recognition.

Lee et al. (2015) proposed that the mapping consistency between phonology and orthography can be measured at the lexical (e.g., character) and sublexical (e.g.,

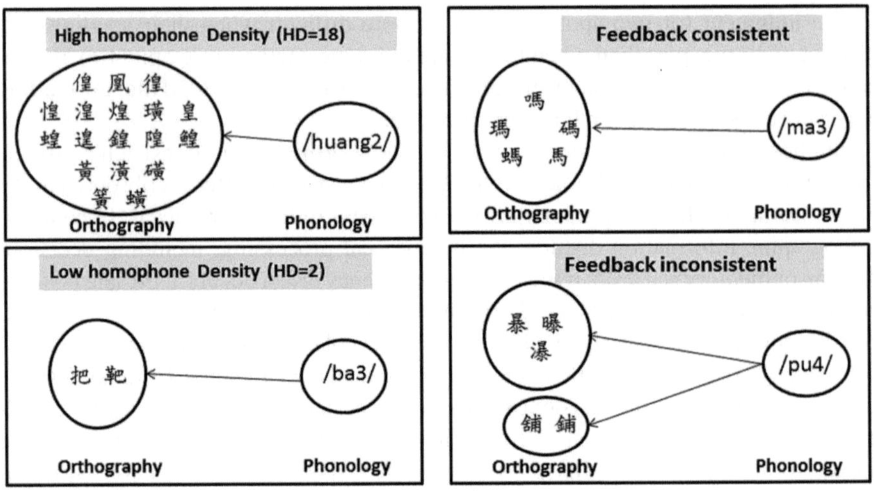

Fig. 2.5 Illustrations of homophone density and P-to-O consistency (feedback consistency) in Chinese

radical) levels (see Fig. 2.5). The first one is the *homophone density*, defined as the number of characters having the same pronunciation (e.g., the syllable cǎi represents these characters as 采, 彩, 採, 綵, 踩, and 睬, and thus its homophone density is six.). The second one is the *P-to-O consistency*, defined as whether a group of homophones can be separated into subgroups by their phonetic radicals. For example, the syllable cǎi is P-to-O consistent since all its homophones contain the same radical 采. By contrast, the syllable chōng is P-to-O inconsistent in that its homophones 沖, 忡, 充, 舂, 衝, and 憧 can be separated into five subsets by their phonetic radicals "中," "充," "臼," "重," and "童."

Chen et al. (2016) utilized ERP to explore the temporal locus of the P-to-O consistency and homophone density effects in the Chinese auditory semantic category task. The P-to-O consistency effect was found at the N400 in 450–650 ms with the frontal-central distribution. Compared to low P-to-O consistency words, high P-to-O consistency words elicited a larger N400 because its orthographic neighbors at the radical level are so abundant that they spread great activation. On the other hand, the homophone density effect appeared at the late positive component (LPC) in 700–900 ms at central-parietal sites. Spoken words with low homophone density were recognized more effortlessly and retrieved without homophone competition, and thus produced a larger LPC than those with high homophone density. Overall, the results confirm the existence of the mapping consistency effect from phonology to orthography on the Chinese spoken language. The sublexical and lexical orthographic variations separately affect the distinct speech processing stages.

Chao et al. (2021) further probed the nature of the P-to-O consistency and homophone density effects by applying event-related fMRI. They used materials similar to those of Chen et al. (2016) in an auditory semantic judgment go/no-go task. The

P-to-O consistency effect within the speech network for phonological processing implied that the sublexical orthographic information modified spoken language by restructuring the nature of phonological representations. In comparison, the homophone density effect in the fusiform area for orthographic access suggested that the lexical orthographic knowledge was activated online during speech processing. The findings verify that these two orthographic effects depend on the different cognitive mechanisms during Chinese spoken word recognition.

The adaptive resonance model indicates that processing spoken language would activate orthographic representations simultaneously, which can modulate phonological processing afterward. (Grainger & Ziegler, 2008; Stone & Van Orden, 1994). Thus, the significance of O-to-P consistency is acknowledged in recognizing spoken words. Chao et al. (2019) found that the homophone density, O-to-P consistency, and character frequency affect participants' performance in Chinese writing-to-dictation tasks. The homophone density has a facilitatory effect on accuracy but an inhibitory effect on generation probability and reaction times. Spoken words with high homophone density exhibited greater production accuracy and longer reaction times. More possible written responses led to a decrease in the generation probability of each word. Notably, the O-to-P consistency effect had a facilitatory effect on the reaction time. Furthermore, Chao et al. (2019) applied ERP to locate an O-to-P consistency-by-homophone density interaction effect on the N400, with subsequent homophone density effects observed on the LPC. Spoken words with high O-to-P consistency elicited larger N400s than those with low P-to-O consistency, but only for syllables with high homophone density. The increased N400s with spoken words with high homophone density and O-to-P consistency resulted from greater numbers of orthographical and phonological neighbors at the sublexical level within the bidirectional mapping. The second-order O-to-P consistency effect in the auditory modality supports the reverberation between orthography and phonology in Chinese spoken word recognition.

2.5 Conclusions

The research on word recognition, mainly from alphabetic languages, has been fruitful. However, languages have inherent differences in their writing systems and spelling rules. An ideal reading model should be universal and mirror typical cognitive processes. It is well known that humans can derive regularities within a stream of visual and auditory stimuli, a fundamental skill crucial for language and reading development. (Chetail, 2015; Frost, 2012; Saffran, 2001). This chapter aims to provide a series of behavioral and neuroimaging evidence to illustrate that, despite Chinese being a non-alphabetic writing system, the bidirectional cross-code consistency effects have been found across Chinese and other writing systems.

More specifically, studies have demonstrated that Chinese readers capture the mapping consistency between character and sound (Fang et al., 1986; Lee et al., 2005; Tzeng et al., 1995). ERP evidence suggests that the feedforward consistency effect in reading Chinese affects early sublexical phonological computation and later lexical semantic competition (Lee et al., 2007; Lee, Tsai, Huang et al., 2006). Moreover, the consistency effect was mainly found in phonograms with large orthographic neighborhoods. This suggests an interplay between orthographic density and the mapping from orthography to phonology in the different stages of lexical processing (Hsu et al., 2009). Characters with dense phonetic combinability are likely to exert more pressure toward awareness of phonological validity represented by phonetic radicals than those with sparse phonetic combinability. Additionally, the neural mechanisms transforming Chinese characters into phonological representations closely resemble those observed in reading alphabetic scripts. (Fiez et al., 1999; Kuo et al., 2003; Lee et al., 2004, 2010; Paulesu et al., 2001). These findings imply that acquiring Chinese characters relies on a statistical mechanism enabling generalization based on similarities.

Two types of feedback consistency effects, namely the phonology-to-orthography consistency (POC) and the homophone density (HD), were also demonstrated in ERP and fMRI studies for Chinese spoken word recognition (Chao et al., 2019, 2021; Chen et al., 2016; Lee et al., 2015). The POC effect on N400 originated from the speech network. It favors the phonological restructuring view and indicates that variations in orthographic elements could potentially reshape existing phonological representations through literacy. The HD effect in the late positive component (LPC) originated from the left inferior frontal gyrus and the temporal-parietal-occipital junction for orthographic processing. The HD effect should be reasoned as the orthographic co-activation. The POC and HD effects reflect different natures of orthographic influence on Chinese spoken word recognition.

These findings suggest that Chinese readers extract information from their linguistic environment regarding the bidirectional consistency between Chinese orthography and phonology and their effect on visual and spoken word recognition. Specifically, the statistical learning mechanism posits that consistent exposure to similar orthographic forms is crucial for recognizing common units within a group of orthographic neighbors. This allows proficient readers to effectively utilize sublexical components, such as phonetic radicals, in establishing efficient mappings from sublexical orthographic units to phonological representations. Above all, this mechanism applies universally across different languages. Despite potential variations in the underlying functional components due to differences in orthographic depth, a similar operational principle is believed to exist across languages.

References

Andrews, S. (1989). Frequency and neighborhood effects on lexical access: Activation or search. *Journal of Experimental Psychology: Learning, Memory, and Cognition, 15*(5), 802–814.

Bolger, D. J., Hornickel, J., Cone, N. E., Burman, D. D., & Booth, J. R. (2008, December). Neural correlates of orthographic and phonological consistency effects in children. *Human Brain Mapping, 29*(12), 1416–1429. https://doi.org/10.1002/hbm.20476

Booth, J. R., Burman, D. D., Meyer, J. R., Gitelman, D. R., Parrish, T. B., & Mesulam, M. M. (2002, May). Functional anatomy of intra- and cross-modal lexical tasks. *Neuroimage, 16*(1), 7–22. https://doi.org/10.1006/nimg.2002.1081

Booth, J. R., Cho, S., Burman, D. D., & Bitan, T. (2007, July). Neural correlates of mapping from phonology to orthography in children performing an auditory spelling task. *Developmental Science, 10*(4), 441–451. https://doi.org/10.1111/j.1467-7687.2007.00598.x

Chang, Y. N., Hsu, C. H., Tsai, J. L., Chen, C. L., & Lee, C. Y. (2016, March). A psycholinguistic database for traditional Chinese character naming. *Behavior Research Methods, 48*(1), 112–122. https://doi.org/10.3758/s13428-014-0559-7

Chao, P. C., Chen, W. F., & Lee, C. Y. (2019). Aug). The second-order effect of orthography-to-phonology mapping consistency on Chinese spoken word recognition. *Journal of Neurolinguistics, 51*, 1–16. https://doi.org/10.1016/j.jneuroling.2018.11.002

Chao, P. C., Chen, W. F., Zevin, J., & Lee, C. Y. (2021, August). Neural correlates of phonology-to-orthography mapping consistency effects on Chinese spoken word recognition. *Brain and Language, 219*, 104961. https://doi.org/10.1016/j.bandl.2021.104961

Chen, W. F., Chao, P. C., Chang, Y. N., Hsu, C. H., & Lee, C. Y. (2016). Jun-Jul). Effects of orthographic consistency and homophone density on Chinese spoken word recognition. *Brain and Language, 157–158*, 51–62. https://doi.org/10.1016/j.bandl.2016.04.005

Chetail, F. (2015). Reconsidering the role of orthographic redundancy in visual word recognition. *Frontiers in Psychology, 6*, 645. https://doi.org/10.3389/fpsyg.2015.00645

Chiu, Y. S., Kuo, W. J., Lee, C. Y., & Tzeng, O. J. L. (2016, January). The explicit and implicit phonological processing of chinese characters and words in Taiwanese deaf signers. *Language and Linguistics, 17*(1), 63–87. https://doi.org/10.1177/1606822x15614518

Ehri, L. C., & Wilce, L. S. (1980). The Influence of Orthography on Readers Conceptualization of the Phonemic Structure of Words. *Applied Psycholinguistics, 1*(4), 371–385. https://doi.org/10.1017/S0142716400009802

Fang, S. P., Horng, R. Y., & Tzeng, O. J. L. (1986). Consistency effects in the Chinese character and pseudo-character naming tasks. *Linguistics, psychology, and the Chinese language*, 11–21.

Fiez, J. A., Balota, D. A., Raichle, M. E., & Petersen, S. E. (1999). Effects of lexicality, frequency, and spelling-to-sound consistency on the functional anatomy of reading. *Neuron, 24*(1), 205–218.

Frost, R. (2012, Oct). Towards a universal model of reading. *Behavioral and Brain Sciences, 35*(5), 263–279. https://doi.org/10.1017/S0140525X11001841

Grainger, J., O'Regan, J. K., Jacobs, A. M., & Segui, J. (1989, March). On the role of competing word units in visual word recognition: the neighborhood frequency effect. *Percept Psychophys, 45*(3), 189–195.

Grainger, J., & Ziegler, J. C. (2008). Cross-code consistency in a functional architecture for word recognition. In E. L. Grigorenko & A. J. Naples (Eds.), *Single-word reading: Behavioral and biological perspectives* (pp. 129–157). Lawrence Erlbaum Associates Publishers.

Hsu, C. H., Lee, C. Y., & Tzeng, O. J. (2014, December). Early MEG markers for reading Chinese phonograms: evidence from radical combinability and consistency effects. *Brain and language, 139*, 1–9. https://doi.org/10.1016/j.bandl.2014.09.008

Hsu, C. H., Tsai, J. L., Lee, C. Y., & Tzeng, O. J. L. (2009, Jan). Orthographic combinability and phonological consistency effects in reading Chinese phonograms: an event-related potential study. *Brain and language, 108*(1), 56–66. https://doi.org/10.1016/j.bandl.2008.09.002

Huang, C. R., & Chen, K. J. (1998). *Academia Sinica balanced corpus*. www.sinica.edu.tw/SinicaCorpus/

Hue, C. W. (1992). Recognition processes in character naming. In H.-C. Chen & O. J. L. Tzeng (Eds.), *Language processing in Chinese. Advances in psychology* (Vol. 90, pp. 93–107). North-Holland. xii, 394 pp.

Jared, D. (1997). Spelling-sound consistency affects the naming of high-frequency words. *Journal of Memory & Language, 36*, 505–529.

Jared, D. (2002). Spelling-sound consistency and regularity effects in word naming. *Journal of Memory and Language, 46*, 723–750.

Kuo, W. J., Yeh, T. C., Lee, C. Y., Wu, Y., Chou, C. C., Ho, L. T., Hung, D. L., Tzeng, O. J., & Hsieh, J. C. (2003, March). Frequency effects of Chinese character processing in the brain: an event-related fMRI study. *Neuroimage, 18*(3), 720–730. https://doi.org/10.1016/s1053-8119(03)00015-6

Lee, C. Y. (2009). The cognitive and neural basis for learning to read Chinese. *Journal of Basic Education, 18*(2), 63–85.

Lee, C. Y., Hsu, C. H., Chang, Y. N., Chen, W. F., & Chao, P. C. (2015, July). The Feedback Consistency Effect in Chinese Character Recognition: Evidence from a Psycholinguistic Norm. *Language and Linguistics, 16*(4), 535–554. https://doi.org/10.1177/1606822x15583238

Lee, C. Y., Huang, H. W., Kuo, W. J., Tsai, J. L., & Tzeng, J. L. O. (2010, January). Cognitive and neural basis of the consistency and lexicality effects in reading Chinese. *Journal of Neurolinguistics, 23*(1), 10–27. https://doi.org/10.1016/j.jneuroling.2009.07.003

Lee, C. Y., Tsai, J. L., Chan, W. H., Hsu, C. H., Hung, D. L., & Tzeng, O. J. (2007, January 22). Temporal dynamics of the consistency effect in reading Chinese: an event-related potentials study. *Neuroreport, 18*(2), 147–151. https://doi.org/10.1097/WNR.0b013e328010d4e4

Lee, C. Y., Tsai, J. L., Chiu, Y. C., Hung, D. L., & Tzeng, O. J. L. (2006). The early extraction of sublexical phonology in reading Chinese pseudocharacters: An event-related potentials study. *Language and Linguistics, 7*(3), 619–636.

Lee, C. Y., Tsai, J. L., Huang, H. W., Hung, D. L., & Tzeng, O. J. (2006, November 22). The temporal signatures of semantic and phonological activations for Chinese sublexical processing: An event-related potential study. *Brain Research, 1121*(1), 150–159. https://doi.org/10.1016/j.brainres.2006.08.117

Lee, C. Y., Tsai, J. L., Kuo, W. J., Yeh, T. C., Wu, Y. T., Ho, L. T., Hung, D. L., Tzeng, O. J., & Hsieh, J. C. (2004, December). Neuronal correlates of consistency and frequency effects on Chinese character naming: an event-related fMRI study [Research Support, Non-U.S. Gov't]. *Neuroimage, 23*(4), 1235–1245. https://doi.org/10.1016/j.neuroimage.2004.07.064

Lee, C. Y., Tsai, J. L., Su, E. C. I., Hung, D. L., & Tzeng, O. J. (2005, December). Consistency, regularity, and frequency effects in naming Chinese characters. *Language and Linguistics, 6*(1), 75–107.

Liberman, A. M. (1992). Chapter 9: The relation of speech to reading and writing. In F. Ram & K. Leonard (Eds.), *Advances in psychology* (Vol. 94, pp. 167–178). North-Holland.

Lo, M., Hue, C.-W., & Tsai, F.-Z. (2007, 12/01). *Chinese readers' knowledge of how chinese orthography represents phonology*, 49.

Montant, M., Schon, D., Anton, J. L., & Ziegler, J. C. (2011). Orthographic Contamination of Broca's Area. *Frontiers in Psychology, 2*, 378. https://doi.org/10.3389/fpsyg.2011.00378

Morais, J., Cary, L., Alegria, J., & Bertelson, P. (1979). Does awareness of speech as a sequence of phones arise spontaneously? *Cognition, 7*(4), 323–331. https://doi.org/10.1016/0010-0277(79)90020-9

Norris, D., McQueen, J. M., & Cutler, A. (2000). Merging information in speech recognition: Feedback is never necessary. *Behavioral and Brain Sciences, 23*(3), 299–325.

Pattamadilok, C., Perre, L., Dufau, S., & Ziegler, J. C. (2009, January). On-line orthographic influences on spoken language in a semantic task. *Journal of Cognitive Neuroscience, 21*(1), 169–179. https://doi.org/10.1162/jocn.2009.21014

Paulesu, E., Demonet, J. F., Fazio, F., McCrory, E., Chanoine, V., Brunswick, N., Cappa, S. F., Cossu, G., Habib, M., Frith, C. D., & Frith, U. (2001). Dyslexia: Cultural diversity and biological unity. *Science, 291*, 2165–2167.

Perfetti, C. A., Cao, F., & Booth, J. (2013, January 1). Specialization and Universals in the Development of Reading Skill: How Chinese Research Informs a Universal Science of Reading. *Scientific Studies of Reading, 17*(1), 5–21. https://doi.org/10.1080/10888438.2012.689786

Perfetti, C. A., & Harris, L. N. (2013, 2013/10/01). Universal reading processes are modulated by language and writing system. *Language Learning and Development, 9*(4), 296–316. https://doi.org/10.1080/15475441.2013.813828

Perre, L., & Ziegler, J. C. (2008). Jan 10). On-line activation of orthography in spoken word recognition. *Brain Research, 1188*, 132–138. https://doi.org/10.1016/j.brainres.2007.10.084

Saffran, J. R. (2001, September). Words in a sea of sounds: the output of infant statistical learning. *Cognition, 81*(2), 149–169. http://www.ncbi.nlm.nih.gov/pubmed/11376640

Seidenberg, M. S. (1985). The time course of phonological code activation in two writing systems. *Cognition, 19*(1), 1–30.

Seidenberg, M. S., & McClelland, J. L. (1989). A distributed, developmental model of word recognition and naming. *Psychological Review, 96*, 523–568.

Seidenberg, M. S., & Tanenhaus, M. K. (1979). Orthographic effects on rhyme monitoring. *Journal of Experimental Psychology-Human Learning and Memory, 5*(6), 546–554. https://doi.org/10.1037/0278-7393.5.6.546

Stone, G. O., & Van Orden, G. C. (1994, December). Building a resonance framework for word recognition using design and system principles. *Journal of Experimental Psychology-Human Perception and Performance, 20*(6), 1248–1268. https://doi.org/10.1037/0096-1523.20.6.1248

Stone, G. O., Vanhoy, M., & Van Orden, G. C. (1997). Perception is a two-street: Feedforward and feedback phonology in visual word recognition. *Journal of Memory & Language, 36*(3), 337–359.

Taraban, R., & McClelland, J. L. (1987). Conspiracy effects in word pronunciation. *Journal of Memory & Language, 26*(6), 608–631.

Tsai, J. L., Lee, C. Y., Tzeng, O. J. L., Hung, D. L., & Yen, N. S. (2004, November). Use of phonological codes for Chinese characters: Evidence from processing of parafoveal preview when reading sentences. *Brain and language, 91*(2), 235–244. https://doi.org/10.1016/j.bandl.2004.02.005

Tzeng, O. J. L., Lin, Z. H., Hung, D. L., & Lee, W. L. (1995). Learning to be a conspirator: A tale of becoming a good Chinese reader. In B. de Gelder & J. Morais (Eds.), *Speech and reading: A comparative approach.* (pp. 227–246). Erlbaum (UK) Taylor & Francis, Publishers.

Tzeng, Y. L., & Lee, C. Y. (2012). The developmental trajectories of orthography-to-phonology mapping consistency in learning to read Chinese. *Contemporary Educational Research Quarterly, 20*(4), 45–84.

Vitevitch, M. S., Luce, P. A., Pisoni, D. B., & Auer, E. T. (1999, June 1–15). Phonotactics, neighborhood activation, and lexical access for spoken words. *Brain and Language, 68*(1–2), 306–311. https://doi.org/10.1006/brln.1999.2116

Weiss, Y., Cweigenberg, H. G., & Booth, J. R. (2018, November). Neural specialization of phonological and semantic processing in young children. *Human Brain Mapping, 39*(11), 4334–4348. https://doi.org/10.1002/hbm.24274

Yang, J., McCandliss, B. D., Shu, H., & Zevin, J. D. (2009). Simulating language-specific and language-general effects in a statistical learning model of Chinese reading. *Journal of Memory and Language, 61*(2), 238–257.

Yoncheva, Y. N., Zevin, J. D., Maurer, U., & McCandliss, B. D. (2010, March). Auditory selective attention to speech modulates activity in the visual word form area. *Cerebral Cortex, 20*(3), 622–632. https://doi.org/10.1093/cercor/bhp129

Ziegler, J. C., & Goswami, U. (2005, January). Reading acquisition, developmental dyslexia, and skilled reading across languages: a psycholinguistic grain size theory. *Psychol Bull, 131*(1), 3–29. https://doi.org/10.1037/0033-2909.131.1.3

Ziegler, J. C., Montant, M., & Jacobs, A. M. (1997). The feedback consistency effect in lexical decision and naming. *Journal of Memory and Language, 37*, 533–554.

Chapter 3
A Visual Approach to the Acquisition of Chinese Characters by the Deaf in Taiwan

James H.-Y. Tai

Abstract A visual approach to the acquisition of Chinese characters is proposed to account for the fact some of the Deaf in Taiwan manage to overcome the cumbersome Chinese characters, thereby to read Chinese with a high level of proficiency. This approach is built up three hitherto little explored threads of thought. Frist, the Chinese characters have a grammar which shares structural principles with signed and spoken languages. Second, both external and internal iconicity in Taiwan Sign Language can be applied to Chinese characters. Third, the Deaf employ visual awareness, rather than phonological awareness, to learn Chinese characters. In conclusion, additional hurdles for the Deaf to read Chinese are briefly sketched in hopes of unraveling the reading ability of Chinese by the Deaf in Taiwan.

Keywords Deaf · Chinese character · Sign language · Iconicity

3.1 Introduction

It has been observed that the reading ability of the Deaf universally lags far behind their hearing peers. The Deaf (with a capital D) refers to the deaf people who use sign language for communication. They generally cannot go beyond the level of the 4th graders in reading comprehension of alphabetic languages such as English (e.g. Quigley & Paul, 1986; Trybus & Karchmer, 1977) and logographic languages such as Chinese (Lin & Li, 1987; Chang, 1987; Tzeng 1996, 1998, 2000). The lack of phonological input and awareness for the Deaf has been identified as the major culprit in causing the problem.

It has been generally assumed that phonological input and awareness are required to decode words in alphabetic languages as well as Chinese characters for the meaning to be incorporated into compounds, phrases, and sentences in reading comprehension. On this assumption, the Chinese logographic writing system would be much harder

J. H.-Y. Tai (✉)
Institute of Linguistics, National Chung Cheng University, Chiayi, Taiwan
e-mail: lngtai@ccu.edu.tw

D. H. Wu (ed.), *Cognitive and Neural Foundations of Chinese Reading*, Chinese Language Learning Sciences, https://doi.org/10.1007/978-981-96-6669-0_3

for the deaf to learn to recognize Chinese characters and to read Chinese texts than alphabetic languages. However, we have noticed that a small proportion of the Deaf population in Taiwan are able to reach the proficiency level to read and write in Chinese.

Unlike the phonological decoding in alphabetical languages, the decoding of Chinese characters is mainly visually oriented. Furthermore, the grammar of Taiwan Sign Language is drastically different from that of Chinese. Therefore, for the Deaf in Taiwan, learning to read Chinese amounts to learning to read a second language without the help of spoken Chinese, as in the case of their hearing peers.

The purpose of this paper is to propose a visual approach to the acquisition of Chinese characters by the Deaf in Taiwan. Our central inquiry is how some of the Deaf manage to overcome the hurdles in learning to read Chinese. This inquiry consists of two parts: recognition of Chinese characters (識字 *shìzì*) and reading Chinese texts (閱讀 *yuèdú*). This paper focusses only on the recognition of Chinese characters by the Deaf in Taiwan, leaving the complicated process of reading Chinese texts for future research.

Regarding the recognition of Chinese characters, three basic theoretical foundations are adopted for this inquiry. First, the Chinese characters have a grammar that shares structural principles with both signed and spoken languages (Myers, 2019). Second, despite the abundance of phonetic radicals, Chinese characters are full of iconic motivations. These motivations are similar to the iconic motivations in sign languages and can be applied to the acquisition the Chinese characters by the Deaf. Third, the Deaf use visual awareness, rather than phonological awareness, to process Chinese characters.

3.2 The Nature of Chinese Characters

3.2.1 Iconicity in Chinese Characters

The first systematic study of Chinese characters appeared in 說文解字 *shuōwénjiězì* 'explaining graphs and analyzing characters' compiled by 許慎 *xǔshèn* around 100 AD. The 文 wén refers to simple non-composite characters, and the 字 *zì* to composite characters. This dictionary contains more than 9,000 Chinese characters in the seal script 篆文 *zhuànwén* to be analyzed with respect to their meaning, composition, pronunciation, and etymology. 許慎 *xǔshèn* did not have the knowledge of the scripts inscribed on the oracle bones and tortoise shells in the late Shang dynasty (fourteenth–eleventh centuries BC), known as 「甲骨文」 *jiǎgǔwén*, which were only discovered around the end of nineteenth century. Besides, his analysis and interpretation were also largely influenced by the cosmological view of his time. He proposed six categories of Chinese characters, referred to as 「六書」 *liùshū* 'six writings'. Although later scholars have proposed different categorizations, the six categories have, nonetheless, guided analysis of Chinese characters for two millennia in the tradition of Chinese philology (cf. Qiu, 2000). Therefore, a few words of explanation of the six categories are desirable and necessary for the purpose of this paper.

Of the six categories, four are concerned with the formation of characters. The first two categories 象形 *xiàngxíng* 'resemble form (of objects)' and 指事 *zhǐshì* 'indicate matters' are about non-composite characters. The former category is exemplified by 日 *rì* 'sun' and 月 yuè 'moon' in 說文解字 *shuōwénjiězì*, and the latter, exemplified by 上 *shàng* 'above' and *xià* 'below'. The third category 會意 *huìyì* 'conjoining meanings' and the fourth category 形聲 *xíngshēng* 'form and meaning' are about composite characters. 說文解字 *shuōwénjiězì* exemplifies the third category with 武 *wǔ* 'martial' and 信 *xìn* 'trust', the fourth category with 江 *jiāng* 'river' and 河 *hé* 'river'. The fourth category contains 'semantic-phonetic' composite characters of which the semantic component provides semantic clues, and the phonetic component, phonological cues. This category is by far the largest as well as the most productive category in the developmental course of Chinese characters, covering more than 90 percent of contemporary Chinese characters in use.

The fifth category is 假借 *jiǎjiè* 'loan-borrowing', which is basically a rebus principle as in Sumerian inscription. Instead of listing the original opaque examples given in 說文解字 *shuōwénjiězì*, we illustrate the principle with the character 能 *néng*. The character is originally a pictographic sketch of a bear. Later, it is borrowed to represent the abstract concept 能 *néng* 'to be able to' through the rebus principle. Then, the character 熊 *xióng* 'bear' is created to represent bears by adding four legs to the bottom of 能 *néng*. Interestingly, 熊 *xióng* is borrowed one more time for the onomatopoeia sound '*xióng*' of burning fire in the expression 熊熊烈火 *xióngxióng lièhuǒ* 'raging flames'.

As a matter of fact, most of the characters in the 形聲 *xióngshēng* category were derived from the 假借 *jiǎjiè* in order to overcome the overwhelmingly large number of homographs. The last of the six categories is 轉注 *zhuǎnzhù*, which is concerned with the extended use of a character in meaning through metonymical or metaphorical uses.

Iconicity plays a key role in both the lexicon and syntax of sign languages (Klima & Bellugi, 1979; Liddell, 2003; Mandel, 1977; Sandler & Lillo-Martin, 2006). Taiwan Sign Language, the native language of the Deaf in Taiwan, is no exception (Myers & Tai, 2005). It is significant to find out to what extent the iconicity in Chinese characters can be of help to the recognition of Chinese characters by the Deaf signers.

Heuristically, we can start with Peirce's three kinds of signs in semiotics, namely, iconic, indexical, and symbolic signs. Iconic signs bear some resemblance to their referents. Indexical signs point to the referents by partial or indirect association. Symbolic signs bear an arbitrary relationship to their referents. While spoken languages, being linear in one dimension, are largely symbolic and arbitrary, thus, the well-known Saussure's arbitrary principle of language. Although it has been convincingly pointed out by Haiman (1980, 1983, 1985) and others that iconicity is pervasive in spoken languages, iconicity in general is not acknowledged in spoken languages as it is in sign languages.

The simple non-composite characters of the first category 象形 *xiàngxíng* 'resemble form (of objects)' are iconic signs and pictorial in origin. Even though their conventionalized contemporary forms are not pictorial anymore, their iconic motivations are easier to interpret and perceive, as in the case of 日 *rì* 'sun' and 月 *yuè* 'moon'. The compound character 明 *míng* 'bright' belongs to the third category 會意 *huìyì*, and the 'conjoining meaning' is transparent and mnemonic to learn. Many characters for animals are of the first category, for instance, 象 *xiàng* 'elephant' and 馬 *mǎ* 'horse'. The character 象 *xiàng* was later borrowed through the rebus principle to mean 'resemble' as in 象形 *xiàngxíng* as well as to be used as the phonetic component of the semantic-phonetic character 像 *xiàng* 'portrait'. In the same vein, perhaps, in different diachronic sequences, the character 馬 *mǎ* was adopted as the phonetic component for the character 媽 *mā* 'mother', and as a loan in the expression of 馬馬虎虎 *mǎmǎhūhū*, which does not mean many horses and tigers, but simply mean 'careless'. Reduplication of nouns has an iconic motivation for plurals in Chinese, as in 家家戶戶 *jiājiāhùhù* 'every household'. For the Deaf, the idiom 家家戶戶 *jiājiāhùhù* is easier to grasp than the idiom 馬馬虎虎 *mǎmǎhūhū*. Because the former is semantic-based, but the latter is sound-based, to which the Deaf have no access through spoken Chinese as their hearing peers.

The simple non-composite Chinese characters of the second category 指事 *zhǐshì* 'indicate matters' as exemplified by 上 *shàng* 'above' and 下 *xià* 'below' can further be understood through 'diagrammatic iconicity' in iconicity theory. As pointed out in Haiman (1980), 'Peirce (1932, pp. 247, 277–282) made a crucial distinction between two types of iconicity, which we may call "imagic" and "diagrammatic". An iconic image is a single sign which resembles its referent with respect to some (not necessarily visual) characteristics, an iconic diagram is a systematic arrangement of signs, none of which necessarily resembles its referent, but whose relationships to each other mirror the relationships of their referents (Haiman ibid., p. 315)'. It can be seen that diagrammatic iconicity has the function of isomorphism in mapping reality to language representation. Chinese characters like 上 *shàng* and 下 *xià* exhibit diagrammatic iconicity. They should be very easy for the Deaf to recognize without the sound input of '*shàng*' and '*xià*', as in the case of hearing Chinese children learning to recognize these two characters in the first year of schooling. It should be noted here that Sutton-Spence and Woll (1999, pp. 129–130) propose a distinction between syntactic and topographic space in sign language. While syntactic space is used to express syntactic relations, topographic space reflects the layout of things in the real world. It stands to reason that the Deaf can simply apply their knowledge of topographic space in signing to the recognition of characters in this category.

Now, the character 上 *shàng* and 下 *xià* can be compounded with simple non-composite characters under the general principle of 會意 *huìyì* 'conjoining meanings' as in 忐 *tǎn* and 忑 *tè* to form the expression 忐忑 *tǎntè* 'mentally disturbed'. The two characters have no sound clues at all. The recognition of this compound word needs no sound input for the Deaf as well as for the hearing learners.

形聲字 *xíngshēngzì* of the fourth category, known as phonetic compounds, are treated by Chinese linguists as representing phonological values without iconicity involved. However, the semantic radicals (語意偏旁 *yǔyìpiānpáng*) of these characters are able to provide semantic values. For instance, the semantic radical 扌 is derived from 手 *shǒu* 'hand' and the characters with this radical on the left in general indicate actions with hands, as in 抓 *zhuā* 'to grab' and 扭 *niǔ* 'to twist'. Similarly, the characters with semantic radical 足 *zú* 'foot' have to do with actions with feet, as in 跑 *pǎo* 'to run' and 跳 *tiào* 'to jump'. Therefore, characters in this category are not purely phonological. The semantic radicals in Chinese serve to indicate general semantic meaning, akin to prefixes like 'pre-' and 're-' in English (Myers, 2019).

Some of the semantic radicals not only provide semantic categories, but also diagrammatic iconicity. The semantic radical 囗 *wéi*, which means 'to circumscribe' or 'to surround', is also diagrammatically iconic to the visual relationship, as can be seen in characters such as 園 *yuán* 'garden', 圃 *pǔ* 'a vegetable garden'; 圈 *quān* 'circle' and *juàn* 圈 'pen of domestic animals'; qiú 囚 'prisoner' and 困 *kùn* 'to trap'.

Some composite characters can be diagrammatically iconic, too. For example, the character 盥 *guàn* 'to wash hand or face', which is diagrammatic iconic in each of the four components, correspond to two hand holding the water above the water container. Another example can be seen in the contrast between 裹 *guǒ* 'to wrap' and 裸 *luǒ* 'naked'. The two characters share the same semantic radical 衣 *yī* 'clothes' and phonetic (presumably) component 果 *guǒ* 'fruit'. Note that the semantic radical 衣 *yī* is transformed to 衤 in 裸 *luǒ* as a prefix, but as a split circumfix in 裹 *guǒ*. It appears that diagrammatic iconicity is involved here. While 裹 *guǒ* symbolizes that the body is inside the clothes, 裸 *luǒ* symbolizes that the body is not covered by the clothes.

There are more cases of diagrammatic iconicity in Chinese characters. However, it is more important to point out here that spoken Chinese is loaded with homophones. For example, the word *lì* can have more than 90 meanings, and the resolution to ambiguity largely depends on different characters, e.g. 力 *lì* 'strength', 立 *lì* 'to erect', 利 *lì* 'benefit', 例 *lì* 'example', 麗 *lì* 'beautiful', and 歷 *lì* 'record', etc. Therefore, Chinese characters play an important role in isomorphism in mapping the signal to meaning. And the function of isomorphism hinges upon diagrammatic iconicity.

In short, the commonly accepted thesis that the system of Chinese characters is based on sound and not on visual graphs must be qualified. Furthermore, diagrammatic iconicity is not limited to the whole or partial isomorphism between the schematized visual world and the linguistic structure. As pointed out by Greenberg (1995), in addition to the above external iconicity, there is also internal iconicity, in which the dominant paradigm of a language can serve as a model for the development of linguistic patterns within the same language. The semantic-phonetic category has indeed become the dominant model for new Chinese characters coined for newly discovered metals and chemicals, as numerous composite characters with semantic radicals, for example, 金 *jīn*, a semantic radical for metals, as in 鉀 *jiǎ* 'potassium', 鉛 *qiān* 'lead', and 鈾 *yòu* 'uranium', etc. Another example is 酉 *yǒu*, a semantic radical for chemicals, as in 酮 *tóng* 'ketone', 酯 *zhǐ* 'ester', and 酶 *méi* 'enzyme', etc.

As a matter of fact, internal iconicity also applies to language contact. A good case in point is the character sign used in Taiwan Sign Language (TSL). Ann (1998) has identified the following methods of constructing character signs in TSL.

1. To imitate the shape of either the whole or a part of a Chinese character by means of handshape: 中， 王， 田
2. To trace either the whole or a part of a Chinese character in the air:千， 就
3. To combine both (1) and (2) methods: 丁， 毛， 河
4. To use both handshape and the mouth: 中， 品
5. To combine a natural sign with tracing: 太

In his study of lexical formation of family names and place names in Taiwan Sign Language, Chang (2011a, pp. 114–142; 2011b, pp. 162–183) has identified more than 300 character signs used in full or partial forms. Chinese character signs are also used in Japan and China, adopting the simplified characters and Japanese characters (Fischer & Gong, 2011). Chinese character signs are on a par with finger spellings of the sign languages based on alphabets, such as in American Sign Language and French Sign Language. While in the sense of 'imitation' or 'modelling', both finger spellings and character signs involve internal iconicity, only character signs can involve both external and internal iconicity, as in the character signs 中 *zhōng* 'middle', 王 *wáng* 'king', and 田 *tián* 'rice field'.

In short, the Chinese character system exhibits both external and internal iconicity. It stands to reason to assume that the Deaf in Taiwan can apply imagic and diagrammatic iconicity in the visual world and sign languages to the learning of Chinese characters.

3.2.2 Grammatical Structures

Chinese characters form a system and are amenable to structural analysis within the framework of modern and contemporary linguistic theory. Wang (1983) and subsequently Myers (2019) have clearly shown that Chinese characters have a rich grammar containing phonetics, phonology, morphology, and syntax akin to natural spoken and signed languages. One of the most fundamental properties of human language is 'duality of patterning', in which the meaningless simplest elements can be put together through 'minimal contrast' to form meaningful words. For example, the meaningless phoneme /o/、/p/、/t/、/s/ can form different words such as 'stop', 'spot', 'tops', 'pots', and 'opts' in English. Similarly, the meaningful Chinese characters can be formed with meaningless simple strokes by means of 'minimal contrast', as illustrated in (1) (Tai, 2021:16 with rearrangements).

(1)
a. contrast with the stroke 點 (、) 'dot'
「大/太/犬」；「几/凡」；「王/玉」；「刀/刃」
b. contrast with the stroke 豎 (丨) 'vertical'
「曰/田/由/甲/申」；「爿/片」 (反方向對照)

c. contrast with the stroke 横 (一) ‘horizontal’
「一/十」；「日/旦」；「不/丕」；「口/曰」
d. contrast with the stroke 撇 (丿) ‘throw away’
「十/千」；「天/矢/夫/失」；「心/必」；「代/伐」

In signed languages, meaningful signs are composed by meaningless elements such as handshapes, locations, movements, orientations coupled with facial expressions. They can be illustrated by the following minimal pairs in Taiwan Sign Language (TSL) (Tai & Tsay, 2021: 348).

(2)
a. Handshape contrast: handshape L  vs. handshape Open A  (same location: chin)

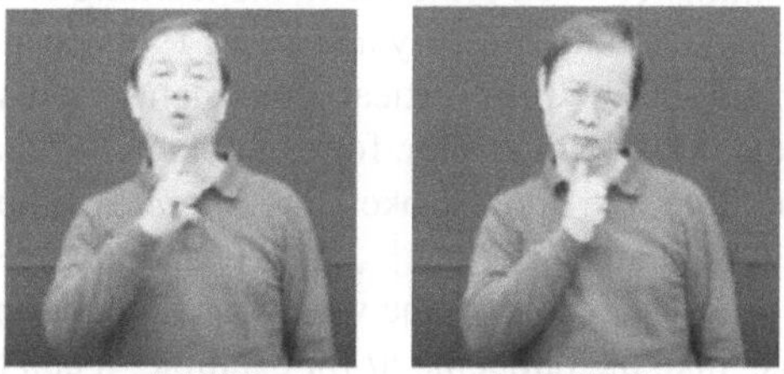

a. RICH b. POOR

b. Location contrast: chest vs. arm (same handshape: Crossed Thumb-Index )

c. CHRISTIAN d. NURSE

c. Movement contrast: downward vs. lateral (same handshape and location)

e. BLACK f. REGRET

d. Orientation contrast: palms down vs. palms facing each other (same handshape and location)

g. NOW h. PERIOD

The Deaf signers in Taiwan or elsewhere in the world should be able to apply their knowledge of the minimal contrast principle to the learning of Chinese characters.

Furthermore, the same stroke can vary according to its position in the character. For example, the stroke [丨] 豎 *shù* 'vertical' consistently takes the form of [丿] 彎 *wān* 'bend' on the left side of a character, for example in the characters 「川」 and 「月」. At the same time, the hook stroke [亅] 鉤 *gōu* will appear on the right of the 「月」, as well as in characters 「句」, 「丁」 and 「事」, etc. Variations of a stroke in different positions are akin to the variations (i.e. allophones) of a phoneme in different positions. Take the phoneme /t/ for example, it can have five allophones as [th]op (top), s[t]op (stop), bu[ʔ]on (button), wri[D]er (writer), and cu[t̚] (cut). Therefore, strokes have the status of 'graphemes' with 'allographs' as variants in character compositions. Signs in sign languages naturally adjust their forms in the discourse as the words will adjust their phonemes in the flow of speech. Therefore, the Deaf can transfer this kind of linguistic knowledge to the recognition of Chinese characters.

The grammar of Chinese characters also has a rich component of morphology, in addition to phonetic and phonological components (Myers, 2019). First, affixation constitutes a central part of morphology. Significantly, Chinese characters exhibit a system akin to affixation in spoken and sign languages. Thus, the semantic radicals derived from the full independent characters in (3) can be treated as affixes in Chinese characters. First, they cannot be independent characters (獨體字 *dútǐzì*) and can only be components (部件 *bùjiàn*) of characters. Second, their semantic categorizations are often broader than the original characters from which they are derived.

(3)
人 → 亻；刀 → 刂; 水 → 氵；艸 → 艹
手 → 扌；心 → 忄；火 → 灬; 衣 → 衤

It is significant to note that while words in spoken language are one-dimensional, characters are two-dimensional with horizontal and vertical axes. On the horizontal axis, some affixes can be placed on the left of a character, as the semantic radical [亻] in 「仁」 and 「他」, thus on a par with a prefix in spoken language. They can also be placed on the right, as the semantical radical [刂], akin to a suffix in spoken languages in 「刺」 and 「刮」 or infixes as [丿] in 「辨」 and [糸] in 「辮」.

On the vertical axis, Chinese characters can also have top-affixes and bottom-affixes. For example, the semantic radical [艹] in 「草」 and 「菜」; and the semantic radical [灬] in 「煎」 and 「煮」. More interestingly, the independent full character 「衣」, in addition to its derived semantic radical [衤] on the left in 「衫」,「袖」, and others, 「衣」 can be split as in 「衷」、「裹」, akin to merged compound words in spoken and sign languages.

In addition to affixation, compounding is also a productive word formation mechanism in the morphology of spoken and signed languages. Chinese characters are loaded with compound characters as illustrated in (4).

(4)
日 + 月 ➔ 明 (semantic radical: 日)
木 + 才 ➔ 材 (semantic radical: 木)
田 + 力➔男 (semantic radical: 田)
耳 + 心➔恥 (semantic radical: 心)
上 + 心➔忐 (semantic radical: 心)
下 + 心➔忑 (semantic radical: 心)

Note that these compound characters cannot be written as two characters with space in between. They must be collapsed into one single square in different ways as illustrated in (4). These examples also illustrate the effect of two-dimensionality of compounding, different from one dimensionality in spoken languages, as 'blue + bird ➔ bluebird', and 'green + house ➔ greenhouse' in English. Otherwise, they are parallel in both meaning change and prosodic adjustment. In the case of Chinese characters, it is the adjustment of the relative proportion of the space in one single square.

Reduplication is another component of morphology. It is very common in both spoken and sign languages to use reduplication to signify plurality, an iconic motivation in both modalities of language. Reduplication with the same motivation can also be illustrated with the familiar example in (5).

(5)
木 + 木 = 林; 木 + 木 + 木 = 森
火 + 火 = 炎; 火 + 火 + 火 = 焱
魚 + 魚 = 䲆; 魚 + 魚 + 魚 = 鱻 (variant character of 「鮮」)

In short, the morphology of Chinese characters is largely similar to that of spoken and sign languages. The profound Deaf can certainly transfer their knowledge of reduplication in sign language to the recognition of Chinese characters.

Myers (2019) has only slightly touched on the syntax of Chinese characters. In this paper, we propose to start with their composition of components within characters, especially the patterns of spatial arrangement of semantic radicals, phonetic radicals, and other kinds of components. The main patterns can be seen in (6).

(6)
1. ⿰ (兩個部件由左至右): 「相」、「肝」、「邦」、「朝」
2. ⿱ (兩個部件由上至下): 「杏」、「吉」、「岩」、「召」

3. ⿲ (三個部件由左至右):「衍」、「班」、「辨」、「瑯」
4. ⿳ (三個部件由上至下):「京」、「岔」、「莫」、「高」
5. ⿴ (兩個部件由外而內):「回」、「因」、「國」、「困」
6. ⿵ (三面包圍–下方開口):「凰」、「問」、「周」、「風」
7. ⿶ (三面包圍–上方開口):「凶」、「幽」、「函」、「凼」
8. ⿷ (三面包圍–右方開口):「匠」、「區」、「巨」、「匹」
9. ⿸ (兩面包圍–由左上至右下):「屠」、「屍」、「屬」
10. ⿹ (兩面包圍–由右上至左下):「包」、「病」、「載」
11. ⿺ (兩面包圍–由左下至右上):「超」、「過、「建」
12. ⿻ (同樣部件的重複):「林」,「森」;「日」、「昌」、「晶」

The 12 patterns of spatial arrangements listed above are based on Unicode 3.0. Other permissible patterns remain to be uncovered.

Furthermore, in many cases, the same semantic element in a character can be structured either as a semantic radical or as a part of the semantic compound. It can be illustrated with the three most productive semantic elements, 'heart', 'hand', and 'foot' as in (7), (8), and (9).

(7) 心 *xīn* 'heart' as independent characters
1. as semantic radical「忄」on the left (152 characters)

忙、忖、忭、忸、伉、快、忻、忮、忪、忱、忡、忤、忨、怖、怕、怦、怛、
怓、怩、怔、怜、怪、恆、怙、怯、怊、怵、怍、怡、怏、恫、恬、恡、恑、
恪、恇、恨、恢、恍、恰、恓、恤、恂、恃、悖、悌、悧、悃、悍、悔、悁、
性、悄、悛、悚、悒、悞、悅、悟、惜、悻、惆、惓、悱、悼、惦、惇、惝、
惕、**惧**、悾、惚、惛、悸、悽、情、悵、惙、悴、悰、惟、惋、惘、愎、愊、
愐、惰、惱、愣、惶、愀、愜、愓、惺、惴、愞、惻、愕、愔、愉、惲、慄、
愷、愾、愧、慌、慊、慉、愴、慎、慷、慍、慢、慟、慣、慨、慷、慳、慴、
慥、慚、慘、慪、慵、憫、憚、憐、憒、憍、憬、憔、憧、憎、憯、憤、懂、
憹、懍、懈、憸、懊、懌、憶、懦、懤、懵、懶、懷、懺、懽、懼、懾。

2. as a part of compound placed at the bottom (82 characters)

忐、忑、忌、志、忍、忘、忿、念、忽、忠、怠、怹、怒、急、怎、思、怨、
恁、恧、恐、恚、恝、息、恕、恣、恩、恙、您、患、悉、悤、悠、恿、悲、
悳、惄、惑、惠、惎、惡、悶、愍、感、愆、想、愁、惹、意、愚、態、慈、
愬、慇、愿、慂、慝、慮、慧、慼、慤、慫、慰、慾、憊、憋、憑、憝、憨、
憩、憙、憲、憖、懋、懇、應、懑、懟、懲、懸、戁、戀、戇。

3. other non-productive patterns of compounding

(a) 愛、慶、憂; (b) 恥 、 懿 、 忒; (c) 忝、恭、慕; (d) 必

It is worthy noting that the anti-ambiguity constraint works here to keep the distinction between the pairs: 忘 ≠ 忙; 忠 ≠ 忡; 怎 ≠ 怍; 愁 ≠ 愀;意 ≠ 愔; 怒 ≠ 怓 悲 ≠ 悱. The anti-ambiguity constraint can be treated as a corollary of the principle of isomorphism for the mapping between meaning and form.

(8) 手 *shǒu* 'hand' as an independent character
1. as a semantic radical「扌」on the left (312 characters)

扎、扒、扑、打、扔、托、扛、扣、扞、扦、扠、把、扳、扮、抔、批、扶、抖、扭、抗、技、抉、抵、折、抓、找、扯、抄、抒、扼、抑、抆、拋、拔、抱、扮、拂、拍、披、拖、拚、抨、抹、抿、拇、拊、抵、拓、拗、拈、拉、拎、拐、拒、拘、招、拄、拙、拆、抽、抻、押、拼、挑、括、拱、拷、拮、指、拯、拽、持、拾、拭、拶、拴、按、挖、挓、挽、捌、捕、挺、捏、挪、捍、挾、捄、挶、捔、捐、捃、振、捉、捎、捘、挫、挱、挨、挹、捂、捭、排、掊、捧、捫、捯、掉、掂、掇、掏、探、掭、推、捺、捻、捩、掄、掠、掆、掛、掯、控、掎、接、捷、据、掬、掘、掐、掮、捡、掀、掙、捵、捨、授、掞、捽、採、措、掃、掩、捲、捱、掖、掗、提、換、揮、揸、搜、握、揚、插、援、揭、揀、揉、揪、揶、掿、揢、揩、揣、揍、揄、捶、揖、摒、揆、揳、揞、揠、揎、撝、揕、掾、揜、搖、損、搬、搞、搇、搶、搭、搏、搗、搓、搔、搪、搥、搵、搐、搧、摀、搽、搾、搠、搌、搆、搉、搤、搳、搕、搘、搢、搦、搰、搨、摁、搊、搋、搛、摸、撧、撤、摘、摔、摧、摺、撇、摟、摞、摳、摜、摑、摏、摻、摶、摽、摭、摛、摴、摋、摠、播、撲、撞、撥、撒、擳、撫、撐、撈、撕、撓、撰、撮、撬、撅、撣、撳、撢、撙、撚、撏、撟、擀、擓、據、擔、擇、擁、擋、擅、撿、撼、擒、擂、擄、撻、撾、擗、擐、抬、擦、擠、擬、擱、擰、擤、擢、擣、擯、擩、擺、擴、擾、擲、擭、擷、擻、攄、擿、攏、攔、攘、攟、攖、攝、攜、攛、攤、攡、攪、攢、攥、攫、攩、攬、攮。

2. as a part of compound placed at the bottom (20 characters)

拿、拏、挈、拳、挲、掌、掣、掔、搴、摩、摹、摯、擧、擊、擎、擘、攀、攣、擪、撆

3. other non-productive patterns of compounding

a. 𢪙、掰; b. 承; c. 拜; 掱

(9) 足 *zú* 'foot' as an independent character.

1. as a semantic radical 「𧾷」 on the left (96 characters)

趴、趵、趺、趿、跂、趾、跋、跛、跑、跗、跌、跎、跅、跏、距、跖、跚、跐、跥、跳、路、跟、跪、跨、跬、跡、跤、趼、跣、跩、跴、踉、跽、蹁、踊、踣、踏、踝、踦、踩、踧、蹁、踶、蹀、跺、蹄、踽、踵、蹅、踹、蹂、踴、蹈、蹋、蹌、蹊、蹉、蹦、蹕、蹣、蹢、蹚、蹟、蹡、蹠、蹧、蹤、蹼、蹯、蹬、蹲、蹻、蹶、蹺、蹭、蹴、躃、躂、躅、躇、躁、躋、躊、躍、躒、躐、躑、躓、躔、躕、躚、躥、躞、躡、躧、躝。

2. as a part of compound placed at the bottom (6 characters)

跫、踅、蹇、躄、蹩、躉

3.2.3 Visual Awareness

A sketch of visual awareness of Chinese characters is proposed below for the profoundly Deaf who use sign language and have little, if any, phonological awareness. The parameters for visual awareness of Chinese characters are listed in (10).

(10)
a. The inventory of simple and complex strokes
b. Composition of strokes (with stroke order in writing)
c. Minimal pairs by strokes and components
d. Semantic components, and no semantic components (including phonetic components for the hearing and those which can be recognized as mere components, for example, 「𦰩」 in the left component of 「難」 and the right component of 「漢」, and 「𢦏」 shared by 「戴」 and 「載」
e. Different patterns of spatial arrangement of strokes and components in squares.
f. knowledge of compound characters and their alternative affixation.

It is the argument of this paper that the profoundly Deaf using Taiwan Sign Language (or other sign languages) can use the visual awareness as sketched in (10) to learn to recognize Chinese characters without phonological inputs.

3.3 Advantage of Deafness to Phonological Noises

First, written Chinese is based on Mandarin Chinese, which is loaded with homophones. For example, 「國語日報辭典」 *guóyǔrìbàocídiǎn* 'Dictionary for the Daily National Language Newspaper, the commonly used Chinese dictionary in Taiwan lists 51 characters under */lì/*, e.g. 力, 立, 利, 歷, 麗…

The hearing readers learned these characters through their pronunciation. It is inevitable that the abundance of homophones would interfere with the processing of Chinese characters. Yet, the Deaf readers can avoid the interference.

Second, Tone 2 and Tone 3 in Mandarin are often not clearly distinguished from each other. So are /l/ and /n/, e.g. 聾人 *lóngrén* 'deaf' and 農人 *nóngrén* 'farmer'. The Deaf only need to recognized different characters to avoid the confounding of the two tones.

Third, there exist regularity and consistency problems. Both involve the distorted sound patterns of phonetic radicals for historical reasons. Mainly, phonetic radicals remain unchanged in form regardless of sound changes through centuries. The regularity problem can be illustrated in (11) and consistency in (12)

(11)
regularity: 昔 */xí/*, 錯 */cuò/*, 借 */jiè/*

(12)
consistency: 由 /*yóu*/: 10 regular (油, 鈾, 蚰, 柚…); 2 irregular (抽 /*chōu*/), 迪 /*dí*/), thus consistency rate: 2/12=0.17

Furthermore, both regularity and consistency are compounded by the frequency effect (Lee et al., 2005).

For the hearing, the errors are exhibited not only in comprehension but also in the production of Chinese characters. However, the deaf can also make errors due to confusion of the similarity between and among characters. It can be illustrated by Chinese characters produced by the Deaf in (13) (Shen, 2023, p. 62).

(13)

1. 構 (溝) 通	2. 薪 (新) 手	3. 偶 (遇) 到
4. 平版 (板)	5. 桃 (挑) 選	6. 遭 (糟) 糕
7. 辨 (辦) 露營	8. 彩卷 (券)	9. 屏束 (東)
10. 主特 (持)	11. 中搶 (槍)	12. 翻釋(譯)

3.4 Advantages of the Deaf in Visual Working and Short Memory

Although the Deaf lack auditory working and short memory in reading, they are compensated by the stronger visual working and short memory than those of the hearing readers (Maller, 2003; Marschark, 2003; Ronnberg, 2003). Recently, Lee (2021) used eye-tracking and ERP to compare the reading processes of English by the Deaf and the hearing, and found that the Deaf pay more attention to the reading than the hearing. It is reasonable to infer that the Deaf in Taiwan can focus more on the forms and meanings of Chinese characters, ignoring their phonology and the ensured irregularity and inconsistency of sound patterns in their mental lexicon.

Flaherty (2003) summarized results from previous experimental studies on visual memory tests of deaf signers and showed that deaf signers consistently outscored their hearing non-signing counterparts. She therefore proposed that long-term use of a sign language may enhance visual abilities and extended her proposal to argue that the long-term use of logographic script Chinese characters appears to boost visual memory. Both learning a sign language and mastering Chinese characters involve a relatively complex task of visual recognition and memory.

Furthermore, in their study comparing deaf signers with and without spoken language base in Taiwan, Chiu and Wu (2013) found that phonological coding is not required for Chinese reading comprehension of the deaf signers without spoken language base. Chiu and Wu (2016) further confirmed that deaf signers use TSL phonological representations during reading.

3.5 Conclusion

In this paper, we propose a visual approach to the acquisition and recognition of Chinese characters by the Deaf in Taiwan. The long-term goal is to develop a fairly elaborate theory, which aims to account for the fact that a certain portion of the profoundly deaf can manage to read Chinese and to write their MA thesis in Chinese. In fact, it has been pointed out by Goldin-Meadow and Mayberry (2001) that reading requires two related but separable capabilities: familiarity with a language and understanding of the mapping between the language and the printed words. The Deaf signers in Taiwan are certainly not only familiar with but skillful in TSL. The mapping between TSL and Chinese characters is facilitated by the iconic motivations and structural similarities between the two.

On the other hand, at this point, we need to have some tentative explanations for the fact that the reading ability of the Deaf universally lags behind their hearing peers, and that many of them cannot go beyond the level of the 4th graders in reading comprehension. First, deaf individuals with little language experience in early life perform poorly in language learning (Mayberry et al., 2002). Note that more than 90% of the deaf are born to hearing parents, thus lack the language environment to develop language like their hearing peers. Between the hearing parents and the deaf child, the communication channels are simple gestures and home signs. The full-fledged sign language is to be developed among the deaf children community when they enter the school for the deaf. Second, the written language is largely based on spoken language, and for the deaf to learn to read amounts to learning a second language because of grammatical differences between signed and spoken languages. Third, without the phonological bases, the Deaf are naturally double-disadvantaged in learning to read as their hearing peers. In the theory of teaching schoolers in elementary schools, the first three grades are focussed on 'learn to read', and from grades four to six, the focus is on 'read to learn'. When the education system fails to teach deaf children to read from the first to the third grade, the majority of deaf children will have difficulties in entering the stage of 'read to learn'. Without the successful second stage, their learning is seriously jeopardized, and, as a consequence, their metacognition and their desire to read for new knowledge are greatly hampered.

The decoding of Chinese characters is the first step towards the comprehension of reading Chinese. The structure of spoken Chinese also needs to be learned for the second step of reading Chinese. This second step consists of the segmentation of Chinese compound words with different structures, and different types of garden paths in processing Chinese simple and complex sentences. Thereby, a theory of the acquisition of Chinese characters and reading largely on the visual basis is proposed to account for the ability to read and write Chinese by the Deaf in Taiwan.

As to the acquisition of Chinese reading, we can start with SVR (simple theory of reading, as expounded by Gough & Tunmer, 1986; Perfetti, 1999; Scarborough, 2001; Tunmer & Hoover, 1992) and the essence of Chinese grammar as identified by Tai (2023). We will step-by-step unweave the entangled threads of reading Chinese by the Deaf and hopefully gradually contribute to the reading education of Chinese by the Deaf children as well as the hearing children with reading difficulty.

References

Ann, J. (1998). Contact between a sign language and a written language: Character signs in Taiwan Sign Language. In C. Lucas (Ed.), *Pinky extension and Eye Gaze: Language use in deaf communities* (pp. 59–99). Gallaudet University Press.

Chang, Bey-Lin (張蓓莉). (1987). Huíguī zhǔliú tīngjué zhàngài xuéshēn gyǔwén nénglì zhī yánjiù 回歸主流聽覺障礙學生語文能力之研究. *Bulletin of Special Education, 3*, 119–134.

Chang, Jung-hsing. 2011a. *Lexical formation of family names in Taiwan sign language*. Craine.

Chang, Jung-hsing. 2011b. *Lexical formation of Place names in Taiwan sign language*. Craine.

Chiu, Yi-Shiuan (邱倚璿), & Ming-Da Wu (吳銘達). (2013). Zhōngwén zìxíng yǔ yǔyīn zhōubiān yùshì xiàoyì: yú jiàojiā yuèdú nénglì shīcōngzhě zhī zhèngjù中文字形與語音周邊預視效益:於較佳閱讀能力失聰者之證據(The parafoveal preview benefits of Chinese orthography and phonology among deaf readers with high reading abilities). *Bulletin of Special Education, 38*(3): 31–54.

Chiu, Y.-S., & Ming-Da, Wu. (2016). Use of phonological representations of Taiwan sign language in Chinese reading: Evidence from deaf signers. *Bulletin of Special Education, 41*(1), 91–109.

Fischer, S., & Gong, Q. (2011). Marked hand configurations in Asian sign languages. In R. Channon & H. van der Hulst (Eds.), *Formational units in sign languages* (pp. 19–42). De Gruyter Mouton.

Flaherty, M. (2003). Sign language and Chinese characters on visual-spatial memory: A literature review. *Perceptual and Motor Skills, 97*(3), 797–802.

Goldin-Meadow, S., & Mayberry, R. I. (2001). How do profoundly deaf children learn to read? *Learning Disabilities Research & Practice, 16*(4), 222–229.

Gough, P. B., & Tunmer, W. E. (1986). Decoding, reading, and reading disability. *Remedial and Special Education, 7*(1), 6–10.

Greenberg, J. H. (1995). On language internal iconicity. In M. Landsberg (Ed.), *Syntactic iconicity and linguistic freezes* (pp. 57–64). Mouton de Gruyter.

Haiman, J. (1980). The iconicity of grammar: Isomorphism and motivation. *Language, 56*(3), 515–540.

Haiman, J. (1983). Iconic and economic motivation. *Language, 59*(4), 781–819.

Haiman, J. (Ed.). (1985). *Iconicity in syntax*. John Benjamins.

Klima, E. S., & Bellugi, U. (1979). *The signs of language*. Harvard University Press.

Lee, C.-L., Hung, D. L., Tse, J.-P., Lee, C.-Y., Tsai, J.-L., & Tzeng, O.-L. (2005). Processing of disyllabic compound words in Chinese aphasia: Evidence for processing limitation hypothesis. *Brain and Language, 92*(2), 168–184.

Lee, B. A. (2021). *Word representation and processing in deaf readers: Evidence from ERPs and eye-tracking* (Ph.D. thesis). San Diego State University, State of California.

Liddell, S. K. (2003). *Grammar, gesture, and meaning in American sign language*. Cambridge University Press.

Lin, Bao-gui (林寶貴), and Zhen-Xian Li (李真賢). (1987). Tīngjué zhàngài xuéshēng guóyǔwén nénglì zhī yánjiù 聽覺障礙學生國語文能力之研究. *Jiàoyù xuéyuàn xuébào* 教育學院學報 *12*, 1–29.

Maller, Susan J. (2003). Intellectual assessment of deaf people: A critical review of core concepts and issues. In *the Oxford handbook of deaf studies, language, and education*, M. Marschark & P. E. Spencer (pp. 451–463). Oxford University Press.

Mandel, M. (1977). Iconic devices in American sign language. In L. A. Friedman (Ed.), *On the other hand* (pp. 57–107). Academic Press.

Marschark, M. (2003). Cognitive functioning in Deaf. In *the Oxford handbook of deaf studies, language, and education*, M. Marschark & P. E. Spencer (pp. 464–477). Oxford University Press.

Mayberry, R. I., Lock, E., & Kazmi, H. (2002). Linguistic ability and early language exposure. *Nature, 417*(6884), 38–38.

Myers, J. (2019). *The grammar of Chinese characters: Productive knowledge of formal patterns in an orthographic system.* Routledge.

Myers, J., & Tai, James H.-Y. (Eds.). (2005). *Language and linguistics* 6.2. Special Issue on Taiwan Sign Language. Institute of Linguistics, Academia Sinica.

Peirce, C. S. (1932). Division of signs. *Collected Papers of C. S. Peirce, 2*, 134–155.

Perfetti, C. A. (1999). Comprehending written language: A blueprint of the reader. In C. M. Brown & P. Hagoort (Eds.), *The neurocognition of language* (pp. 167–208). Oxford University Press.

Qiu, Xigui. (裘錫圭). (2000). *Chinese Writing* (G. L. Mattos & J. Norman, Trans). Society for the Study of Early China and the Institute of East Asian Studies, University of California.

Quigley, S., & Paul, P. (1986). A perspective on academic achievement. In D. Luterman (Ed.), *Deafness in perspective* (pp. 55–86). Taylor Francis.

Ronnberg, J. (2003). Cognition in the hearing impaired and deaf as a bridge between signal and dialogue: A framework and a model. *International Journal of Audiology, 42*, S68–S76.

Sandler, W., & Lillo-Martin, D. (2006). *Sign language and linguistic universals.* Cambridge University Press.

Scarborough, H. S. (2001). Connecting early language and literacy to later reading (dis)abilities: Evidence, theory, and practice. In S. B. Neuman & D. K. Dickinson (Eds.), *Handbook of early literacy research* (pp. 97–110). Guilford Press.

Shen, Jian-Wei. (沈建偉). (2023). *Táiwān tīngzhàngzhě zhōngwén shūmiànyǔ piānwù fēnxī* 臺灣聽障者中文書面語偏誤分析 (An Analysis of Errors in Written Chinese in the Hearing-Impaired in Taiwan) (M.A. thesis). National Chung Cheng University, Taiwan.

Sutton-Spence, R., & Woll, B. (1999). *The linguistics of British sign language.* Cambridge University Press.

Tai, James H.-Y. (2013). Reflections on typological characterization of Chinese grammar. *Human Language Resources and Linguistic Typology*, 59–87. Papers from the Fourth International Conference on Sinology. Edited by Chiu-yu Tseng, Academia Sinica.

Tai, James H.-Y. (戴浩一). (2021). Jiànjīyú hànzì yǔ hànyǔ jiégòu de zhōngwén èryǔ yuèdú lǐlùn 建基於漢字與漢語結構的中文二語閱讀理論 (Towards an innovative theory of reading Chinese as a second language). *Chinese Language Learning and Technology, 1*(1), 13–35.

Tai, James H.-Y. (2023). Quest for the essence of Chinese grammar. *Special Issue for the 50th anniversary of Journal of Chinese Linguistics.*

Tai, James H.-Y. (to appear 2025). Iconic Motivations in Sequence in Spoken and Signed Languages. In O. Fischer, K. Akita, & P. Perniss (Eds.), *Handbook on iconicity in language.* Oxford University Press.

Tai, James H.-Y. (戴浩一), & Jane Tsay (蔡素娟). 2021. Shǒuyǔ yǔyánxué 手語語言學 (Sign Linguistics). In Shuan Fan Huang (黃宣範) (Ed.), *Yǔyánxué: jiégòu, rènzhī yǔ wénhuà de tànsuǒ* 語言學: 結構、認知與文化的探索 (Linguistics: An inquiry into structure, cognition and culture) (pp. 343–382). National Taiwan University Press.

Trybus, R. J., & Karchmer, M. A. (1977). School achievement scores of hearing-impaired children: National data on achievement status and growth patterns. *American Annals of the Deaf, 122*, 62–66.

Tunmer, W. E., & Hoover, W. A. (1992). Cognitive and linguistic factors in learning to read. In P. B. Gough, L. C. Ehri, & R. Treiman (Eds.), *Reading acquisition* (pp. 175–214). Erlbaum.

Tzeng, Shi-Jie (曾世杰). (1996). Tīngzhàngzhě yuèdú kùnnánde kěnéng yuányīn zhī yī 聽障者閱讀困難的可能原因之一. *Bulletin of Special Education, 60*, 17–20.

Tzeng, Shi-Jie (曾世杰). (1998). Lóngrén yuèdú zhōngwéncí shí cíhuì chùjiē yǔ duǎnqí jìyì lìchéng zhī zhuǎnlù xiànxiàng yánjiù 聾人閱讀中文詞時詞彙觸接與短期記憶歷程之轉錄現象研究

(Deaf Chinese readers' recoding strategy in the lexical access and short-term retention of Chinese characters). *Bulletin of Special Education and Rehabilitation, 6*, 65–101.

Tzeng, Shi-Jie (曾世杰). (2000). Zhōngwén yuèdú bùxū shēngyùn juézhì de zhèngjù 中文閱讀不需聲韻覺識的證據. Paper presented at the Proceedings of the Special Education Teaching Conference of the 89th Academic Year of National Taitung Normal University, pp. 79–94.

Wang, Jason Chia-Sheng. (1983). *Toward a generative grammar of Chinese character structure and stroke order* (Ph.D. thesis). University of Wisconsin-Madison, Madison.

Part II
Reading Shaped by Predictive and Contextual Information

Chapter 4
Relationship Between Shared Book Reading, Predictive Brain Signals, and Language Development in Infants

Shinmin Wang

Abstract The ability to predict future events based on past experience is crucial for cognitive functions like sensory processing and cognitive development. This chapter explores the relationship between predictive brain signals and language development, and how shared book reading experiences influence both. Predictive brain signals in infants aged 6, 12 and 18 months were measured using functional near-infrared spectroscopy (fNIRS) and the visual-omission paradigm. Strong correlations were found between the strength of these signals and both expressive vocabulary (at 12, 18, 26 and 47 months) and sentence complexity (at 18 and 30 months). Furthermore, shared book reading experiences reported by parents showed a significant correlation with predictive brain signals. This connection mediates the relationship between shared book reading experiences and children's expressive vocabulary. The findings emphasize the important role of early experiences in shaping the neural mechanisms underlying language acquisition and highlight the potential of shared reading to enhance predictive brain activity, thereby fostering language development. However, limitations regarding the parent-reported nature of shared reading assessment warrant further investigation using more objective measures.

Keywords Sensory prediction · fNIRS · Shared reading · Language · Young children

The ability to predict future events based on past experience is essential for various cognitive functions, such as sensory processing and cognitive development. This chapter aims to summarize existing evidence on the relationship between predictive brain signals and language development and the influence of shared book reading experiences on both. The first part of the chapter presents a review of evidence regarding the presence of predictive abilities at the neural level in the developing brains of infants, highlighting the use of the visual-omission paradigm to measure these predictive signals. The second part of the chapter explores the relationship

S. Wang (✉)
Department of Child and Family Science, National Taiwan Normal University, Taipei, Taiwan
e-mail: s.wang@ntnu.edu.tw

D. H. Wu (ed.), *Cognitive and Neural Foundations of Chinese Reading*, Chinese Language Learning Sciences, https://doi.org/10.1007/978-981-96-6669-0_4

between predictive brain signals, as measured using functional near-infrared spectroscopy responses in the occipital lobe during unexpected visual-omission tasks, and language development in infants. Finally, the chapter discusses how regular reading sessions with infants offer constant opportunities for them to engage in predicting forthcoming information within a structured context. Thus, enhanced shared book reading experiences strengthen predictive brain signals in infants, thereby facilitating language acquisition.

4.1 Measurement of Predictive Signals in Infant Brains

4.1.1 Neural Responses Related to Predictions in Infant Brains

Infants are innately capable of top-down sensory prediction; this ability is potentially key to efficient information processing in the brain (Aslin, 2014; Emberson, 2017; Kveraga et al., 2007). Numerous behavioral studies have indicated that infants can form expectations across various perceptual and cognitive domains through brief learning experiences, enabling them to adjust their responses to new environmental stimuli (see Aslin, 2007, 2014 for a review). Recently, an increasing body of research has aimed to unveil the neural mechanisms underlying learning and anticipation in infants by investigating the influence of prior information on neural responses through functional near-infrared spectroscopy (fNIRS) (Wang et al., 2022). fNIRS is a noninvasive optical imaging method employed to monitor functional changes in cortical hemodynamics.

For example, Nakano et al. (2008) used fNIRS and observed distinct responses in the temporal parietal versus frontal regions of three-month-old infants' brains to auditory cues when the infant predicted subsequent auditory stimuli. The auditory cue signaling the presence of a second stimulus triggered more pronounced brain activation in the bilateral temporal parietal regions of the cortex compared with the cue signaling its absence. Similarly, Gervain et al. (2008) employed fNIRS to assess the cortical responses to sound sequences following either a repetitive or nonrepetitive pattern in neonatal brains. Their findings revealed increased activation in response to the repetition pattern, particularly in the left-hemisphere frontal areas, with this difference intensifying with prolonged exposure to the repetition pattern. In addition to fNIRS studies, investigations employing electroencephalography (EEG) have revealed neural responses related to learning and anticipation in infants. For example, Kouider et al. (2015) exposed 12-month-old infants to auditory cues predicting visual stimuli or auditory cues with varying degrees of predictability or not predictive at all. Their findings indicated substantial differences in event-related potentials between the predictive cues and those that were less predictive or not predictive at all.

The three aforementioned studies have provided compelling evidence of brain signals associated with various aspects of prediction, including sensory associations and distributional learning. However, these studies have offered less definitive insight into how predictions regarding future sensory input manifest neurally, because the neural responses revealed in these studies could stem from either prediction errors or nonspecific surprise effects. A paradigm that offers greater clarity in assessing whether predictions can modulate responses in the sensory cortex is the stimulus omission paradigm (Emberson et al., 2015). This paradigm can reveal whether the infant brain harbors the fundamental architecture necessary for supporting sensory prediction.

4.1.2 A Special Case of Prediction: Stimulus Omission

In the stimulus omission paradigm, individuals are presented with pairs of stimuli, such as pairs of auditory tones. Interestingly, Hughes and colleagues (2001) revealed a consistent response in the auditory region of the brain even when the second tone was intentionally withheld from the participant. In simpler terms, the brain had learned to associate these two tones as a pair and expected the second tone to follow. Consequently, when the second tone was absent, the brain maintained its response. This discovery led to a modification of the paradigm designed by Emberson et al. (2015, 2017), wherein the pairs now involve both auditory and visual stimuli.

In a series of studies conducted by Emberson et al., six-month-old infants were exposed to a visual stimulus shortly after a corresponding auditory cue, which induced an association between the sound and a subsequent visual stimulus. Following the acquisition of these cross-modal associations, the infants were subjected to trials in which the auditory cue was presented, but the expected visual stimulus was unexpectedly absent. Through fNIRS, the researchers first verified the presence of activations in the temporal and occipital cortices of full-term 6-month-olds that are related to auditory and visual processing. Subsequently, they assessed occipital activations during trials where visual stimuli were omitted. This unique experimental design featuring visual omission eliminated any ambiguity regarding whether the response to the altered cued stimulus was a reaction to a deviation because any response to the omitted visual stimulus necessarily indicated prediction due to the absence of sensory information.

The findings revealed a strong occipital response during the visual-omission trials, but this response was apparent only after the auditory cue had consistently been paired with the visual stimulus. Consequently, the ability to make predictions based on prior learning is evident in six-month-old infants with typical development. In a subsequent study, severely premature infants underwent testing at 6 months of age (adjusted for gestational age) by using the same visual-omission paradigm (Emberson et al., 2017). The results indicated the absence of occipital responses to the visual-omission trials in these premature infants. This suggests a deficiency in a fundamental aspect of the brain's prediction mechanism among preterm infants, a population at a higher risk of developmental delays across various cognitive domains. Taken together, these studies suggest that the predictive brain signal revealed in the visual-omission paradigm may function as a developmental mechanism and play a crucial role in how the infant brain utilizes experience to facilitate typical developmental trajectories.

4.2 The Predictive Brain and Language Development

Linguistic development is a key part of infant development. Preterm infants have been widely reported to be at a higher risk of experiencing delays in language development compared with their full-term counterparts. Many preterm children demonstrate delays in global delays in both gestural communication and linguistic abilities, which become apparent as early as 12 months of age (e.g., Cattani et al., 2010) and persist by the age of two (e.g., Sansavini et al., 2011). Given the documented absence of robust predictive brain signals in preterm infants at 6 months of age (Emberson et al., 2017) and the prevalence of delayed language development in this population (e.g., Cattani et al., 2010; Sansavini et al., 2011), a connection between the predictive brain signals and language development can be anticipated. To understand this potential link, relevant findings from a series of studies conducted by Wang and colleagues are further reviewed (Wang et al., 2022; Wang et al., 2022).

4.2.1 Word Level

The association between predictive brain signals and language capacity at the word level was investigated in a study by Wang, Tzeng et al. (2022) involving 23 Taiwanese full-term infants aged 12 months. The expressive vocabulary skills of the 12-month-old infants were assessed using the Mandarin Chinese version (Liu & Tsao, 2010) of the MacArthur-Bates Communicative Development Inventory (Taiwan) (MCDI-T), whereas predictive brain signals were measured using fNIRS during the visual-omission task, consistent with the approach used in studies by Emberson et al. (2015, 2017). The predictive brain signal, indicated by oxy-Hb changes during the visual-omission trials, exhibited a significant correlation with infants' expressive vocabulary ($r = 0.599$), as depicted in Fig. 4.1. Further hierarchical regression analysis revealed that the predictive brain signal significantly predicted expressive vocabulary even after controlling for the influence of maternal education and early shared reading experiences. Specifically, it accounted for 13.6% of the variance in expressive vocabulary. These findings support the contention that predictive brain signals are uniquely associated with language development in children.

To further investigate the potential long-term relationship between predictive brain signals and language proficiency at the word level, Wang, Tzeng et al. (2022) examined the association between infants' predictive brain signals and their vocabulary development in a longitudinal study. The strength of the predictive brain signals was determined by the magnitude of the neural response to the unexpected omission of a visual stimulus in the visual-omission task described previously. Vocabulary

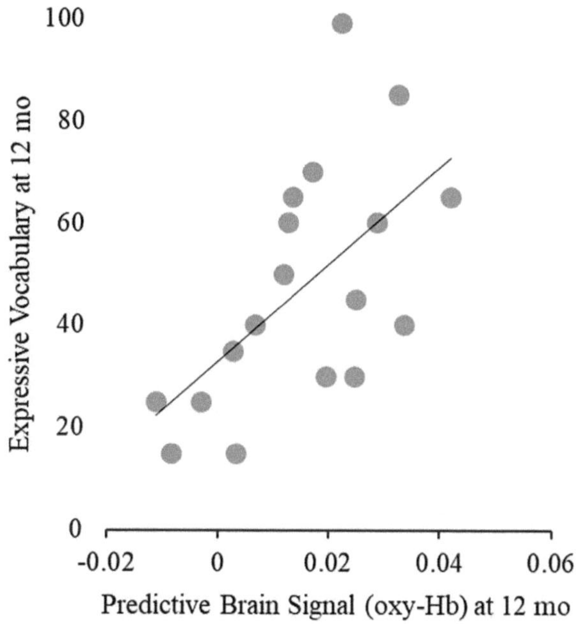

Fig. 4.1 Scatter plot, along with the best-fitting regression line, illustrating the correlations between brain predictive signal (Oxy-Hb) and expressive vocabulary (PR)

scores were assessed using the MCDI-T (Liu & Tsao, 2010) when infants were 12 and 18 months of age. Significant positive correlations were observed between the predictive neural signal at 6 months and MCDI-T expressive vocabulary scores at 12 months (*rho* = 0.454, $p < 0.05$, one-tailed) and 18 months (*rho* = 0.657, $p < 0.01$, one-tailed; Fig. 2a–b).

To extend our investigation, we continued to track the progress of these children as they reached 26, 30, and 47 months of age, respectively. At 26 and 30 months, their vocabulary capacities were estimated again by using the parent-reported MCDI-T. At 47 months, the children were invited to the laboratory, where they underwent specific assessments. These assessments included the Receptive Vocabulary, Information, and Picture Naming subtests from The Wechsler Preschool and Primary Scale of Intelligence (fifth edition), abbreviated as WPPSI-IV. The Receptive Vocabulary subtest evaluates the children's ability to correctly identify spoken words, whereas the Information subtest assesses their general knowledge, long-term memory, and factual knowledge. The Verbal Comprehension Index comprises the Receptive Vocabulary and Information subtests. Additionally, the Picture Naming subtest measures the children's ability to name pictorial stimuli. The Verbal Acquisition Index comprises the Receptive Vocabulary and Picture Naming subtests.

The results revealed significant positive correlations between the predictive neural signal at 6 months and MCDI-T expressive vocabulary scores at 26 months (*rho* = 0.492, $p < 0.05$, one-tailed; Fig. 2c). Remarkably, a similar positive correlation was also observed between total-Hb changes during the visual-omission trials at 6 months and children's verbal acquisition ability, as measured using the WPPSI-IV

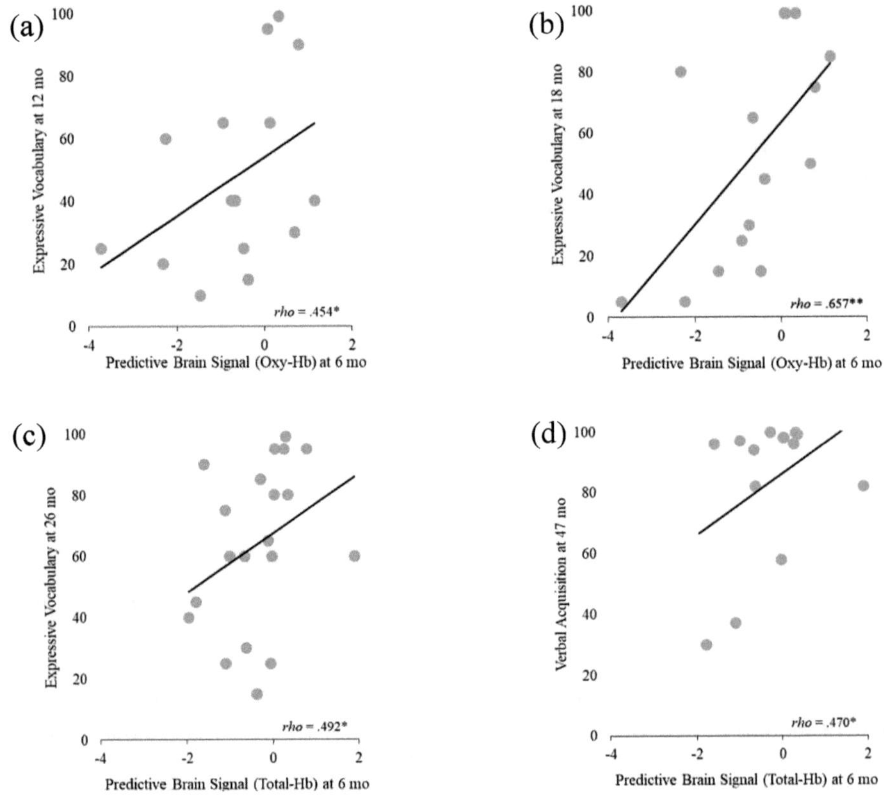

Fig. 4.2 Longitudinal relationships between predictive brain signals at 6 months and **a** expressive vocabulary at 12 months, **b** expressive vocabulary at 18 months, **c** expressive vocabulary at 26 months, and **d** verbal acquisition ability at 47 months. The trend line represents the least squares regression line

at 47 months (*rho* = 0.470, $p < 0.05$, one-tailed; Fig. 2d). These findings jointly indicate the predictive power of the brain's signals at 6 months in forecasting the subsequent development of vocabulary at 12, 18, 26, and 47 months. This robust evidence strongly supports the hypothesis that top-down prediction at the neural level plays a crucial role in language development among children.

4.2.2 Sentence Level

The predictive brain signal was also found to be associated with language development at the sentence level. The level of language proficiency was assessed through parental reports of sentence complexity and word combination skills of children by using the MCDI-T (Liu & Tsao, 2010). The assessment of sentence complexity involved 25 pairs of sentences, with one representing typical uncomplicated grammatical structures and the other demonstrating more complex structures. Parents selected the complexity level of sentences based on their toddlers' current expressive abilities. The word combination section comprised the three longest sentences spoken by the toddler, which were written by parents. Positive correlations were observed between total-Hb changes during the visual-omission trials at 6 months and sentence complexity at 18 months (*rho* = 0.454, $p < 0.05$, one-tailed) as well as word combination skills at 30 months (*rho* = 0.410, $p < 0.05$, one-tailed; Fig. 3a–b). Additionally, positive correlations were observed between total-Hb changes during the visual-omission trials at 18 months and sentence complexity at 30 months (*rho* = 0.434, $p < 0.05$, one-tailed; Fig. 3c).

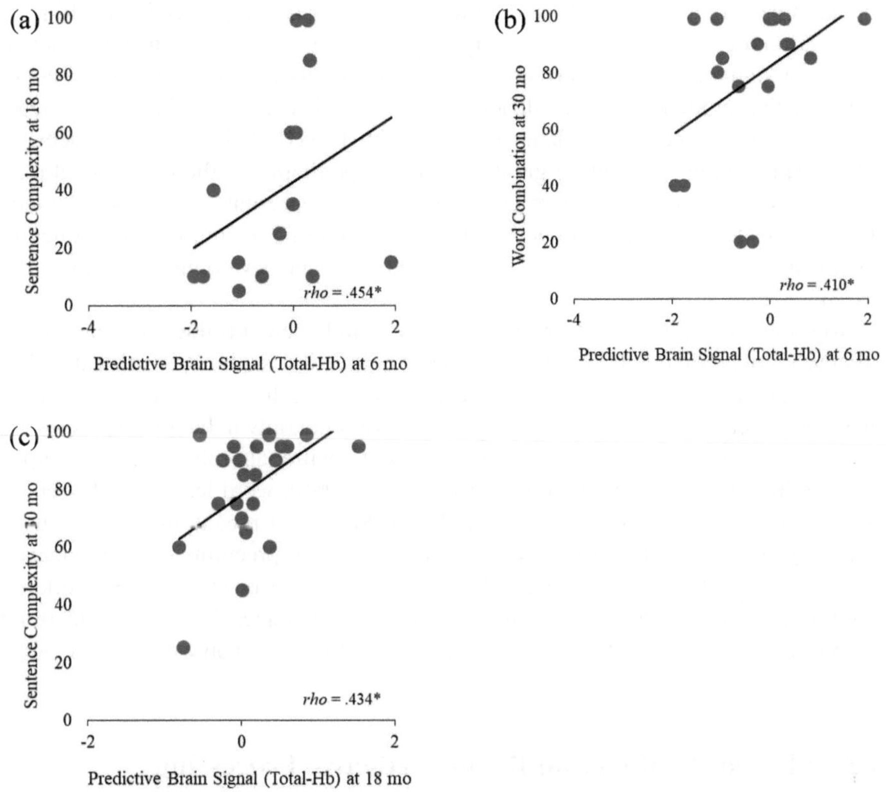

Fig. 4.3 Longitudinal relationships between **a** predictive brain signal at 6 months and sentence complexity at 18 months, **b** predictive brain signal at 6 months and word combination at 30 months, and **c** predictive brain signal at 18 months and sentence complexity at 30 months. The trend line represents the least squares regression line

4.2.3 Prediction Plays a Key Role in Language Development

The longitudinal associations observed between predictive brain signals in infants and various aspects of language abilities in later stages suggest that sensory cortex prediction plays a pivotal role in the initial formation of linguistic representations. Specifically, with regard to word-level proficiency, these top-down sensory prediction signals in the infant brain were consistently correlated with infants' vocabulary production across different stages of early language development. Wang, Zhang et al. (2022) argued that one implication of these findings is the involvement of prediction in the regulation of speech production. During word production, the activation of a motor speech unit prompts the internal model of sensory-motor connections to generate predictions regarding its auditory outcomes, which are then compared with the auditory target. If a match is identified, the motor signal proceeds, resulting in articulation that aligns with the auditory target. However, in the absence of a match, the motor signal undergoes correction until it accurately produces the target sound. Through this predictive and feedback loop, a neural representation of oral motor movement associated with spoken words is established (Hickok, 2012; Hickok et al., 2011). Thus, infants exhibiting greater top-down prediction at the neural level may tend to develop a faster and more precise oral motor representation of the utterance that matches the speaker's production they are attempting to emulate, consequently enhancing their word production in subsequent language development (Wang et al., 2022).

Top-down sensory prediction may serve as a fundamental framework supporting other types of predictions involved in language processing, such as transitional probabilities, statistical preemption, and construction learning. Johnson et al. (2013) argue that these three types of prediction play crucial roles not only in language processing but also in language development. For example, transitional probabilities are implicated in initial word segmentation (Saffran et al., 1996), word learning (Hay et al., 2011), and grammar learning (Saffran, 2002). Statistical preemption can be used to manage overgeneralization, where, for instance, *went* preempts *goed* in contexts where *A* is expected, but *B* is repeatedly used instead (Johnson et al., 2013). Children learn that *B*, not *A*, is the appropriate form to use in language. Construction learning becomes necessary when mapping form and meaning in novel phrasal constructions.

4.3 Shared Reading and Brain Predictive Processing

The aforementioned finding indicates that the predictive brain signal serves as a facilitator of language development of infants (Wang et al., 2022; Wang et al., 2022). One related aspect of interest is the impact of early experiences on both the brain's predictive signal and language development. By the second year of life, a child's experiences are primarily shaped by their parents. Shared book reading is a form of parent–child interaction that offers a consistent and relatively structured context for infants to exercise their predictive abilities (Wang et al., 2022). Wang et al. explored this aspect of development with 12-month-old infants by investigating the contribution of shared book reading experiences to the formation of predictive brain signals. Specifically, the study aimed to ascertain whether predictive brain signals explain the observed concurrent relationship between shared book reading experience and vocabulary development during infancy. Predictive brain signals were again assessed using occipital responses to the unexpected visual-omission trials. Infants' expressive vocabulary was estimated using the MCDI-T (Liu & Tsao, 2010), whereas infants' shared book reading experiences were reported by parents using the infant version of StimQ-READ subscale (Dreyer et al., 1996; Mendelsohn et al., 1999).

Remarkably, infants' shared book reading experiences, as measured using the StimQ-READ subscale, were significantly correlated with both the strength of the predictive brain signal and infants' expressive vocabulary. Further mediation analysis revealed that approximately one-third of the effect of shared reading experiences on expressive vocabulary could be attributed to predictive brain signals in infants, even after maternal education level was controlled for. The comprehensive model explained 59.3% of the variance in the expressive vocabulary of infants (see Fig. 4.4). These findings imply that beyond its direct influence on expressive vocabulary, a richer shared reading experience also enhances predictive brain activity, thereby fostering vocabulary development in infancy (Wang et al., 2022).

4.4 Potential Reasons for Shared Reading Contributing to Predictive Brain Signals

Investigating the contribution of shared book reading experience between parents and infants to the development of robust predictive brain signals is noteworthy. The StimQ-READ subscale assesses factors such as reading frequency (e.g., how often parents read to their child per week), book content (e.g., whether parents read books about daily life to their child), and verbal labeling (e.g., whether parents label pictures while reading to their child). Although the StimQ-READ subscale offers valuable insights into the overall shared reading experience of infants, it provides limited information on how parents interact with their child during reading sessions. However, findings from other studies have elucidated the effects of such interaction. A recent study compared mothers who read books to 6-month-old infants and those who engaged in playing with toys with their infants. The results revealed that mothers used more varied vocabulary and engaged in more verbal interaction with their infants during reading sessions than play sessions, despite the infants' lack of verbal communication skills at 6 months (Wu et al., 2024). Similarly, observational data from another recent study involving 29 parent–infant dyads during shared reading experiences revealed a natural tendency of parents to engage in verbal communication with their infants (Chang et al., 2022). Additionally, evidence indicates that

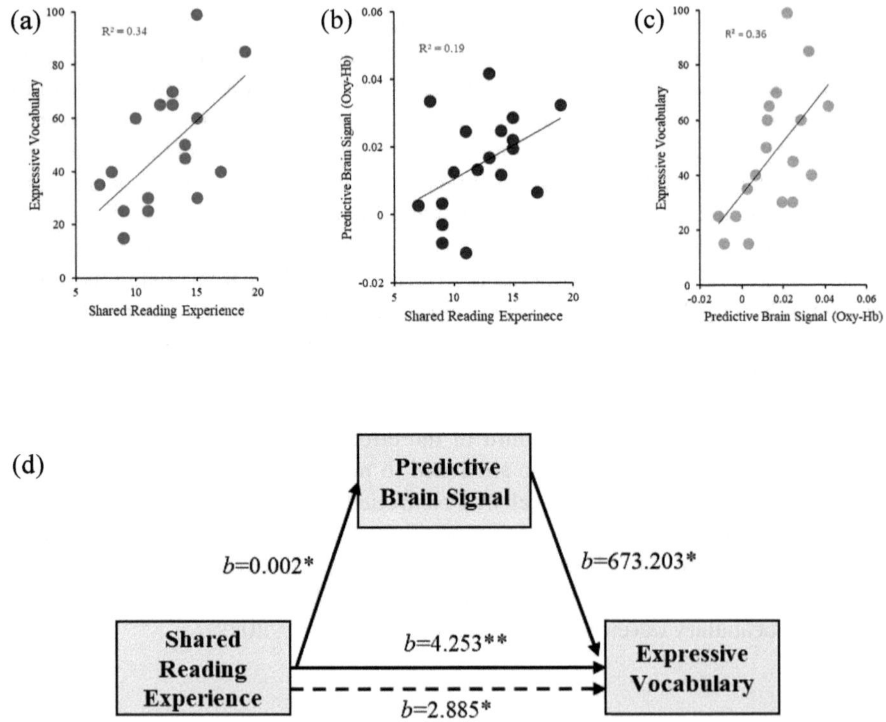

Fig. 4.4 Three scatter plots depicting zero-order correlations, each accompanied by best-fitting regression lines. These plots illustrate the relationships between **a** shared reading experience (raw scores) and expressive vocabulary (PR), **b** shared reading experience (raw scores) and predictive brain signal (Oxy-Hb), and **c** predictive brain signal (Oxy-Hb) and expressive vocabulary (PR). In the mediation model **d**, the influence of shared reading experience on expressive vocabulary is displayed, mediated by the predictive brain signal. The lower path is denoted by solid and dashed arrows, representing results with and without the inclusion of the mediator in the model, respectively. Significant paths are denoted by asterisks ($^{**}p < 0.01$, $^{*}p < 0.05$)

adults spontaneously employ strategies such as prompting children to make predictions and inferences during shared book reading interactions with their infants (e.g., Chang & Huang, 2016; Chang & Luo, 2020), even when the infants have limited spoken language skills (Chang et al., 2022). For example, mothers of 14-month-old infants begin encouraging their infants to predict forthcoming events based on the story being read (Chang & Luo, 2020).

Regular reading sessions with infants and young children offer constant opportunities for them to actively engage in language processing. Language processing involves synthesizing information from a variety of sources, including knowledge about the world, lexical and syntactic cues, and insights into others' mental states (Garnsey et al., 1997; Tanenhaus et al., 1995; Trueswell et al., 1994). These inputs are then used to anticipate how the linguistic signal will unfold (Schrimpf et al., 2021). Extensive behavioral and neural research has demonstrated that listeners employ predictive processing strategies (Brothers et al., 2019; Kuperberg & Jaeger, 2016; Shain et al., 2020; Smith & Levy, 2013). Consequently, children exposed to frequent shared book reading experiences are more likely to engage in predictive processing, thus shaping their predictive brain signals. An intriguing hypothesis is that the strength of predictive brain signals could be further enhanced if parents actively prompt children to make predictions, such as prompting them to anticipate upcoming words (e.g., He was so tired that he wanted to…) or forthcoming events during reading sessions with their children.

The study conducted by Wang, Tzeng et al. (2022) presents preliminary but important evidence suggesting that engaging in shared reading with 12-month-old infants can positively influence the development of their predictive brain signals. Infants who participate in more frequent shared reading activities exhibit stronger predictive brain signals during the visual-omission task. Although this study provides valuable insights, it has certain limitations associated with the measurement of shared reading experience. Shared reading experience was assessed using the StimQ-READ subscale, a parent-reported instrument. Although the StimQ-READ subscale has been successfully used in other studies on the relationship between shared reading experience and brain functions (Hutton et al., 2015, 2020), it had some shortcomings when applied in the present study. First, its reliance on parental reports makes it susceptible to biases arising from memory inaccuracies or social desirability (i.e., the tendency to present oneself in a favorable light). Second, it does not directly evaluate parental interactions with their 12-month-old infants during shared book reading sessions. Consequently, drawing definitive conclusions regarding the depth of parent–child interactions during shared book reading and its potential impact on the development of predictive brain signals is challenging.

4.5 Conclusion

Predictions play a crucial role in the establishment of linguistic representations. This chapter presents a synthesis of recent empirical studies elucidating how the strength of predictive signals in the infant brain predicts later language abilities at both the word and sentence levels across various stages of early development. Additionally, evidence is provided to suggest that infants who engage in more shared book reading experiences tend to exhibit stronger predictive signals in the brain, consequently enhancing their language acquisition skills. Furthermore, potential explanations are discussed as to why shared book reading experiences may serve as effective facilitators for the development of predictive signals in the infant brain. Specifically, shared book reading offers preverbal infants a structured context to consistently utilize their predictive abilities through language processing. Parental scaffolding, such as encouraging infants to predict forthcoming events based on the story being read, further enhances this process. Although the notion that shared book reading fosters predictive ability at the neural level among infants is compelling, evidence for such benefits is limited by the weaknesses of the present study, particularly regarding the measurement techniques used to assess shared book reading experiences. Future studies should, in particular, elucidate the mechanisms linking shared reading with language development and the development of predictive signals in the infant brain.

References

Aslin, R. N. (2007). What's in a look? *Developmental Science, 10*(1), 48–53. https://doi.org/10.1111/j.1467-7687.2007.00563.x

Aslin, R. N. (2014). Infant learning: Historical, conceptual, and methodological challenges. *Infancy, 19*(1), 2–27. https://doi.org/10.1111/infa.12036

Brothers, T., Dave, S., Hoversten, L. J., Traxler, M. J., & Swaab, T. Y. (2019). Flexible predictions during listening comprehension: Speaker reliability affects anticipatory processes. *Neuropsychologia, 135*, Article 107225. https://doi.org/10.1016/j.neuropsychologia.2019.107225

Cattani, A., Bonifacio, S., Fertz, M., Iverson, J. M., Zocconi, E., & Caselli, M. C. (2010). Communicative and linguistic development in preterm children: A longitudinal study from 12 to 24 months. *International Journal of Language & Communication Disorders, 45*(2), 162–173. https://doi.org/10.3109/13682820902818870

Chang, C.-J., & Huang, C.-C. (2016). Mother–child talk during joint book reading in two social classes in Taiwan: Interaction strategies and information types. *Applied Psycholinguistics, 37*(2), 387–410. https://doi.org/10.1017/S0142716415000041

Chang, C.-J., & Luo, Y.-H. (2020). A longitudinal study of maternal interaction strategies during joint book-reading in Taiwan. *Journal of Child Language, 47*(2), 401–417. https://doi.org/10.1017/S0305000919000746

Chang, C. S., Hsieh, F.-J., Chen, T.-Y., Wu, S. C., Tzeng, O. J. L., & Wang, S. (2022). Revisiting dialogic reading strategies with 12-month-old infants. *Early Childhood Education Journal.* https://doi.org/10.1007/s10643-022-01385-4

Dreyer, B. P., Mendelsohn, A. L., & Tamis-LeMonda, C. S. (1996). Assessing the child's cognitive home environment through parental report: Reliability and validity. *Early Development and Parenting, 5*(4), 271–287. https://doi.org/10.1002/(SICI)1099-0917(199612)5:4%3c271::AID-EDP138%3e3.0.CO;2-D

Emberson, L. L. (2017). How does experience shape early development? Considering the role of top-down mechanisms. *Advances in Child Development and Behavior, 52*, 1–41. https://doi.org/10.1016/bs.acdb.2016.10.001

Emberson, L. L., Boldin, A. M., Riccio, J. E., Guillet, R., & Aslin, R. N. (2017). Deficits in top-down sensory prediction in infants at risk due to premature birth. *Current Biology, 27*(3), 431–436. https://doi.org/10.1016/j.cub.2016.12.028

Emberson, L. L., Richards, J. E., & Aslin, R. N. (2015). Top-down modulation in the infant brain : Learning-induced expectations rapidly affect the sensory cortex at 6 months. *Proceedings of the National Academy of Sciences of the United States of America, 112*(31), 9585–9590. https://doi.org/10.1073/pnas.1510343112

Garnsey, S. M., Pearlmutter, N. J., Myers, E., & Lotocky, M. A. (1997). The contributions of verb bias and plausibility to the comprehension of temporarily ambiguous sentences. *Journal of Memory and Language, 37*(1), 58–93. https://doi.org/10.1006/jmla.1997.2512

Gervain, J., Macagno, F., Cogoi, S., Peña, M., & Mehler, J. (2008). The neonate brain detects speech structure. *Proceedings of the National Academy of Sciences, 105*(37), 14222–14227. https://doi.org/10.1073/pnas.0806530105

Hay, J. F., Pelucchi, B., Estes, K. G., & Saffran, J. R. (2011). Linking sounds to meanings: Infant statistical learning in a natural language. *Cognitive Psychology, 63*(2), 93–106. https://doi.org/10.1016/j.cogpsych.2011.06.002

Hickok, G. (2012). The cortical organization of speech processing: Feedback control and predictive coding the context of a dual-stream model. *Journal of Communication Disorders, 45*(6), 393–402. https://doi.org/10.1016/j.jcomdis.2012.06.004

Hickok, G., Houde, J., & Rong, F. (2011). Sensorimotor integration in speech processing: Computational basis and neural organization. *Neuron, 69*(3), 407–422. https://doi.org/10.1016/j.neuron.2011.01.019

Hughes, H. C., Darcey, T. M., Barkan, H. I., Williamson, P. D., Roberts, D. W., & Aslin, C. H. (2001). Responses of human auditory association cortex to the omission of an expected acoustic event. *NeuroImage, 13*(6), 1073–1089. https://doi.org/10.1006/nimg.2001.0766

Hutton, J. S., Dudley, J., Horowitz-Kraus, T., DeWitt, T., & Holland, S. K. (2020). Associations between home literacy environment, brain white matter integrity and cognitive abilities in preschool-age children. *Acta Paediatrica, 109*(7), 1376–1386. https://doi.org/10.1111/apa.15124

Hutton, J. S., Horowitz-Kraus, T., DeWitt, T., Holland, S. K., & Mendelsohn, A. L. (2015). Home reading environment and brain activation in preschool children listening to stories. *Pediatrics, 136*(3), 466–478. https://doi.org/10.1542/peds.2015-0359

Johnson, M. A., Turk-Browne, N. B., & Goldberg, A. E. (2013). Prediction plays a key role in language development as well as processing. *Behavioral and Brain Sciences, 36*(4), 360–361. https://doi.org/10.1017/S0140525X12002609

Kouider, S., Long, B., Le Stanc, L., Charron, S., Fievet, A.-C., Barbosa, L. S., & Gelskov, S. V. (2015). Neural dynamics of prediction and surprise in infants. In (Vol. 6).

Kuperberg, G. R., & Jaeger, T. F. (2016). What do we mean by prediction in language comprehension? *Language, Cognition and Neuroscience, 31*(1), 32–59. https://doi.org/10.1080/23273798.2015.1102299

Kveraga, K., Ghuman, A. S., & Bar, M. (2007). Top-down predictions in the cognitive brain. *Brain and Cognition, 65*(2), 145–168. https://doi.org/10.1016/j.bandc.2007.06.007

Liu, H.-M., & Tsao, F. M. (2010). The standardized and application of Mandarin-Chinese communicative developmental inventory for infants and toddlers. *Formosa Journal of Mental Health, 23*(4), 503–534. https://doi.org/10.30074/FJMH

Mendelsohn, A., Dreyer, B., Tamis-LeMonda, C., & Ahuja, P. (1999). Validity of stimQ, a scale for assessing the cognitive home environment. *Journal of Developmental & Behavioral Pediatrics, 20*(5), 399.

Nakano, T., Homae, F., Watanabe, H., & Taga, G. (2008). Anticipatory cortical activation precedes auditory events in sleeping infants. *PLoS ONE, 3*(12), e3912–e3912. https://doi.org/10.1371/journal.pone.0003912

Saffran, J. R. (2002). Constraints on statistical language learning. *Journal of Memory and Language, 47*(1), 172–196. https://doi.org/10.1006/jmla.2001.2839

Saffran, J. R., Aslin, R. N., & Newport, E. L. (1996). Statistical learning by 8-month-old infants. *Science, 274*(5294), 1926–1928. https://doi.org/10.1126/science.274.5294.192

Sansavini, A., Guarini, A., Savini, S., Broccoli, S., Justice, L., Alessandroni, R., & Faldella, G. (2011). Longitudinal trajectories of gestural and linguistic abilities in very preterm infants in the second year of life. *Neuropsychologia, 49*(13), 3677–3688. https://doi.org/10.1016/j.neuropsychologia.2011.09.023

Schrimpf, M., Blank, I. A., Tuckute, G., Kauf, C., Hosseini, E. A., Kanwisher, N., Tenenbaum, J. B., & Fedorenko, E. (2021). The neural architecture of language: Integrative modeling converges on predictive processing. *Proceedings of the National Academy of Sciences, 118*(45), Article e2105646118. https://doi.org/10.1073/pnas.2105646118

Shain, C., Blank, I. A., van Schijndel, M., Schuler, W., & Fedorenko, E. (2020). FMRI reveals language-specific predictive coding during naturalistic sentence comprehension. *Neuropsychologia, 138*, Article 107307. https://doi.org/10.1016/j.neuropsychologia.2019.107307

Smith, N. J., & Levy, R. (2013). The effect of word predictability on reading time is logarithmic. *Cognition, 128*(3), 302–319. https://doi.org/10.1016/j.cognition.2013.02.013

Tanenhaus, M. K., Spivey-Knowlton, M. J., Eberhard, K. M., & Sedivy, J. C. (1995). Integration of visual and linguistic information in spoken language comprehension. *Science, 268*(5217), 1632–1634. https://doi.org/10.1126/science.7777863

Trueswell, J. C., Tanenhaus, M. K., & Garnsey, S. M. (1994). Semantic influences on parsing: Use of thematic role information in syntactic ambiguity resolution. *Journal of Memory and Language, 33*(3), 285–318. https://doi.org/10.1006/jmla.1994.1014

Wang, S., Tzeng, O. J. L., & Aslin, R. N. (2022). Predictive brain signals mediate association between shared reading and expressive vocabulary in infants. *PLoS ONE*. https://doi.org/10.1371/journal.pone.0272438

Wang, S., Zhang, X., Hong, T., Tzeng, O. J. L., & Aslin, R. N. (2022). Top-down sensory prediction in the infant brain at 6 months is correlated with language development at 12 and 18 months. *Brain and Language, 230*, 105–129. https://doi.org/10.1016/j.bandl.2022.105129

Wu, S.-C., Tzeng, O. J. L., & Wang, S. (2024). Mothers used wider vocabulary and talked to their six-month-old infants more during shared book reading than when they played with toys. *Acta Paediatrica, 113*(1), 84–90. https://doi.org/10.1111/apa.17004

Chapter 5
Evidence of Eye Movements for the Role of Contextual Predictability in Chinese Word Recognition During the Reading of Sentences

Jie-Li Tsai

Abstract When reading sentences, as more words are read, the context is built up and updated from preceding and current word meanings. Contextual processing interacting with lexical processing play a crucial role in reading proficiency for both sentence comprehension and word recognition. In this chapter, the issue of how contextual information affects the lexical processing of an upcoming word in Chinese sentence reading is addressed. For alphabetic languages, eye movement studies consistently showed the influence of contextual constraint on word recognition. The fixation duration on words, which were predictable from the preceding context, was shorter than on the unpredictable words. For reading Chinese, the use of contextual information could be more essential than in alphabetic languages due to the ambiguities of word boundaries and meanings in Chinese sentences. This chapter introduce studies of manipulating word predictability and word frequency to highlight the involvement of contextual information on Chinese word recognition in sentence reading.

Keywords Word predictability · Word frequency · Eye movements · Chinese reading

5.1 Introduction

As we read a sentence, the visual and lexical properties of words are retrieved through a sequence of eye fixations. With each word recognized in sequence, the context is progressively constructed by assimilating the meanings of preceding words. While there is general agreement that contextual information is employed for semantic

J.-L. Tsai (✉)
Department of Psychology, National Chengchi University, Taipei, Taiwan
e-mail: jltsai@nccu.edu.tw

Research Center for Mind, Brain, and Learning, National Chengchi University, Taipei, Taiwan

D. H. Wu (ed.), *Cognitive and Neural Foundations of Chinese Reading*, Chinese Language Learning Sciences, https://doi.org/10.1007/978-981-96-6669-0_5

integration of sentence meaning, there is still no consensus on whether contextual processing takes place concurrently with or prior to lexical processing. The central inquiries revolve around whether top-down contextual constraints can influence the bottom-up encoding of words and at what point in the process that these contextual effects may occur during sentence reading.

Among the research which investigates the mechanism of how sentence constraint is involved in word processing, one of the approaches in experimental design is to manipulate word predictability in a sentence reading task. Researchers often employ a cloze task to establish a word's predictability based on the preceding sentence fragment. In a pioneering study by Stanovich and West (1983), an interaction effect between word predictability and word frequency in word naming was observed. The results indicated a more pronounced facilitative effect of word predictability for low-frequency words compared to high-frequency words. These findings suggest that contextual predictability can indeed improve the efficiency of lexical processing. However, the evidence obtained in a naming task with isolated word presentation may not be generalized to sentence reading that occurs in daily reading.

On-line recording of eye movements during reading provides an opportunity to reveal how contextual information influences reading processes before, during, or after the reader's eyes fixate on words in a sentence. Previous studies have demonstrated that contextual predictability affected fixation durations and probabilities of words in sentence reading (Balota et al., 1985; Ehrlich & Rayner, 1981; Inhoff, 1984; Rayner & Well, 1996). If a word is predictable from the preceding context, the first-pass durations on the word were shorter than when it appeared in the unpredictable context. The findings suggest that contextual information may be involved during or before lexical processing. Moreover, the evidence of predictable words exhibited a higher skipping rate than unpredictable words implies that the influence of word predictability can occur early in lexical processing while a word is parafoveally previewed.

The essential issue is whether the intrinsic word frequency's lexical activation can be influenced by extrinsic contextual predictability. Models of eye movement control during reading, such as E-Z Reader and SWIFT, highlight the significance of lexical processing and utilize word frequency and predictability as crucial factors. The computational simulations of both models have successfully predicted the effects of eye movement behaviors in reading experiments (e.g., Engbert et al., 2005; Reichle et al., 2003). In E-Z Reader, lexical processing involves two stages: the first stage L1 starts preliminary lexical processing of the fixated word and ends to trigger planning of the next eye movement. The second stage L2 completes the processing of the word to induce the shift of attention to the next word. Word frequency and predictability are pivotal in triggering oculomotor or attentional actions in both stages. Previous eye movement studies have manipulated word frequency and predictability and consistently showed their independent effects on fixation durations (Altarriba et al., 1996; Gollan et al., 2011; Miellet et al., 2007; Rayner et al., 2001, 2004, 2006). These findings had also been replicated with an additive simulation model with the E-Z Reader using word frequency and predictability in the lexical processing stages (Rayner et al., 2004).

Another issue regarding whether the context can affect early lexical processing for a parafoveally previewed word can be examined by word skipping. In E-Z Reader, word skipping occurs when lexical processing is complete and saccade programming to the next word begins, while in SWIFT, it happens when the word's activation field is too low to be chosen as the saccade target. Both models suggest that word predictability affects word lexical activation when parafoveally previewed. Many studies consistently showed higher skipping rates for predictable words than for unpredictable words (see Brysbaert & Vitu, 1998, for a review). Calvo and Meseguer (2002) explored the influence of intrinsic and extrinsic factors on eye movement measures, demonstrating the effect of contextual predictability on fixation duration and regression rate, with unpredictable words resulting in longer fixations and higher regression rates. Kliegl et al. (2004) analyzed eye movement data from the Potsdam Sentence Corpus, which has 144 sentences read by 65 participants. They found the effects of word length, frequency, and predictability on fixation durations and probabilities for both early and late measures. The results suggest that contextual constraints not only aid subsequent text integration but also influence earlier reading processes, leading to word skipping.

Word skipping for predictable words could be due to a guess strategy of the next word according to the strong contextual constraint, or the context directly facilitates lexical processing of the parafoveally previewed word. There were studies manipulating word frequency and predictability to examine the hypothesis that early contextual predictability could influence lexical activation of the previewed words in parafovea. Rayner et al. (2004) observed a higher rate of word skipping for predictable, high-frequency words compared to low-frequency or unpredictable words. They suggested that unpredictable words might be identified in parafovea and saccadic programming so the next word is initiated. However, the difficulty of an unpredictable word to be integrated into the preceding sentence's meaning leads to the immediate cancellation of the planning of skipping the word. Consequently, the post-lexical integration of a word's meaning can swiftly impact its skipping rate. Recent research examined the interaction of word frequency and word predictability on duration measures and took the distance from the space before the target to the preceding pre-target fixation, i.e., launch distance, into account (Hand et al., 2010). The results showed the significant interaction effects of word frequency and word predictability with different patterns at the Near (1 to 3 characters) and Middle (4 to 6 characters) launch sites. At the Near launch site, the word predictability effect was more pronounced for low-frequency words; whereas at the Middle launch site, the effect was stronger for high-frequency words. Hand et al. (2010) suggested that the quality of parafoveal preview, as indicated by launch distance, should be integrated into eye movement models to elucidate the underlying reading process. However, reconciling the contrasting patterns of the interaction between word predictability and frequency at the Near and Middle launch sites within a single process of parafoveal preview in the reading model remains challenging.

In the contemporary Chinese writing system, sentences are constructed horizontally with strings of characters arranged from left to right. Chinese characters are considered perceptual units due to their spatial organization and linguistic role. In terms of spatial layout, each character occupies a rectangular area of uniform size, with spaces of equal width separating them regardless of whether they belong to the same word or different words. Characters are composed of component radicals, which can be decomposed into elementary strokes. The elements of a character are contained within a compact space, giving them the appearance of visually integrated and isolated objects (Tsai & McConkie, 2003). Functionally, Chinese characters serve as the fundamental written elements of the spoken language, typically corresponding to morphemes and syllables in the modern system. At the word level, characters are the constituents of Chinese words. According to the Chinese word corpus of Academia Sinica Taiwan (2004), the majority of words (76%) consist of two or three characters. In terms of token frequency, single-character words account for 54% of occurrences, while two-character words account for 42% (Yen et al., 2009). Words can be monomorphemic or polymorphemic, and the number of morphemes within a word corresponds to the number of characters in most situations. There are some exceptions that some two-character words are monomorphemic. Many characters can function independently as single words or as part of compound words. Some characters may be specific to one word, while others may be used in multiple words. Nevertheless, the meanings of characters as individual words or morphemes are not always directly related to the meanings of the words containing them, leading to potential ambiguities in word parsing during reading. Unlike alphabetic language scripts, Chinese lacks visual spacing between words, which makes word segmentation and meaning ambiguity resolution more challenging in sentence processing (Tsai, 2014). As a result, contextual information may be essential for effectively disambiguating word boundaries and meanings when reading Chinese sentences.

Previous eye movement studies of reading Chinese had manipulated word frequency or word predictability. Chinese readers tend to skip high-frequency words more often and spend less time fixating on them compared to low-frequency words (Tsai et al., 2006; Yan et al., 2006; Yang & McConkie, 1999). The effect size of word frequency in Chinese resembles that observed in the English studies (Yan et al., 2006). In terms of contextual effects, Rayner et al. (2005) adopted an experimental design similar to that of Rayner and Well (1996), manipulating the predictability of target words based on prior context into high, medium, or low predictability conditions. Their findings showed that Chinese readers exhibited comparable eye movement patterns to English readers, with longer gaze durations and lower skipping rates for low-predictability words compared to medium and high predictability words. This suggests that either Chinese readers or English readers utilize contextual cues during reading. However, the pattern of word skipping differed from that observed by Rayner and Well (1996) among English readers, who tended to skip words primarily when they were highly predictable. A possible explanation for such discrepancy is that Chinese characters are positioned closer to fixation than English before a potential skipping occurs (Rayner et al., 2005). Alternatively, it could imply that Chinese readers may rely more on contextual information than English readers since medium-predictable words can facilitate lexical processing than low-predictable words.

Research has illustrated the interaction between word frequency and predictability in affecting fixation durations and probabilities among English and French readers (Hand et al., 2010; Miellet et al., 2007; Rayner et al., 2004). In the case of Chinese readers, the absence of explicit word segmentation and the character-to-word orthography suggests that word predictability might have a more pronounced impact than word frequency. Understanding the effects of word frequency and predictability can provide insights into how both inherent lexical familiarity and external contextual predictability influence eye movement control during Chinese sentence reading.

5.2 Methodology

In this chapter, we presented a study conducted by Tsai et al. (2009) manipulating the word frequency and predictability of target words within Chinese sentences. We provided methodology and results details to exemplify a standard eye movement experiment concerning reading, incorporating the manipulation of word predictability and frequency. The study recruited thirty-two college and graduate students who were native Chinese speakers from Taiwan, compensated for their participation. Participants were tasked to read sentences in traditional Chinese, with 40% of these sentences followed by comprehension questions. For each trial, a sentence was displayed in a single line in the middle of the computer screen. Each character and its preceding space subtends 0.9 degrees of visual angle. Eye movements during sentence reading were recorded using an SR Research EyeLink 1000 eye tracker, operating at a sampling rate of 1000 Hz.

The frequency and contextual predictability of the target word were manipulated using a 2 by 2 factorial design. A total of one hundred two-character words were selected as target words from a Chinese corpus (2004). Among these, half were designated as high-frequency (HF) words, and the other half as low-frequency (LF) words. The average word frequency per million was 91.25 for high-frequency target words and 1.59 for low-frequency target words (Table 5.1). Two hundred sentences containing either 25 or 26 characters were generated, with a target word placed at positions 11 to 16 within each sentence. For each target word, two sentences were created. In one sentence, the target word could be predicted from the preceding context, while in the other sentence, the target word was unpredictable based on the context. Target predictability was evaluated by a group of twenty participants who were not involved in the eye movement experiment of reading. These participants were presented with sentence fragments preceding the target words and asked to provide the word that first came to mind to complete each fragment. The predictability value was calculated based on the proportion of participants who filled in the target word as their first choice. The mean cloze probability value was 0.80 for high-frequency predictable target words, 0.75 for low-frequency predictable target words, and 0.01 for both high-frequency and low-frequency unpredictable targets. Four lists were devised to balance the predictability and frequency of the target word in each sentence.

Table 5.1 Means of word frequency and predictability of target words and example sentences for each condition

Condition		Word Frequency	Word Predictability	Example Sentences
HF	P	9..25	0.80	這個風景名勝區每年都吸引許多遊客到此來欣賞美麗景緻 This scenic area attracts many **tourists** every year to enjoy the beautiful scenery
	UP	91.25	0.01	小時候常去的古厝如今只留下遊客丟棄的垃圾及滿屋的塗鴉 The old house that I used to visit in my childhood is now left with garbage thrown away by **tourists** and graffiti all over the house
LF	P	1.59	0.75	亞馬遜河流域因擁有廣大的熱帶雨林而成為生態中的瑰寶 The Amazon River Basin, due to its vast tropical **rainforests**, becomes an ecological gem
	UP	1.59	0.01	林教授貢獻一生的心力在探索雨林的生態及其可用的資源 Prof. Lin has dedicated his life to exploring the ecology of the **rainforest** and its available resources

Note The unit of word frequency is in occurrence per million words, word predictability in percentage. Target words are highlighted with bolds and underlines in the example sentences. HF = high frequency; LF = low frequency; P = predictable; UP = unpredictable

5.3 Results

Four types of fixation durations on the target words were computed. The first is First Fixation Duration (FFD), which represents the duration of the initial fixation on a word during the first pass of reading. Gaze Duration (GD) is the total duration of the first pass fixations on a word before moving to another word. Single Fixation Duration (SFD) denotes the duration of fixation on a word that only receives one fixation during the first pass reading. Total Viewing Time (TVT) encompasses the sum of all fixation durations on a word, including instances of rereading. Additionally, two measures of fixation probabilities were assessed during the initial pass: the rate of skipping the target word and the rate of refixating on the target word. Trials (3.4%) in which participants blinked while fixating on the target or when the target was fixated at the beginning or end of the trial were excluded from analysis. Viewing durations shorter than 80 ms or longer than 800 ms were also excluded (1.3%) from the analysis. Table 5.2 displays the overall means of the fixation durations and probabilities for the target words across each condition. A Linear Mixed Model

(LMM; Baayen, 2008; Baayen et al., 2008) analysis was conducted to examine the effects of word frequency and predictability on each eye movement measure, treating both participants and items as random effects. These analyses were performed using the lmer program from the lme4 package (Bates & Maechler, 2010) in the R system (R Development Core Team, 2010). The estimated effect size (b), standard error (SE), and t-value were reported for each effect.

Target Fixation Durations. The analysis of target fixation durations was conducted based on 2446, 2132, 2435, and 2570 observations for First Fixation Duration (FFD), Single Fixation Duration (SFD), Gaze Duration (GD), and Total Viewing Time (TVT), respectively. Across all measures, target fixation durations were consistently shorter for high-frequency or predictable words compared to low-frequency or unpredictable words (Fig. 5.1). The word frequency effect (high frequency vs. low frequency) was marginally significant for FFD and SFD (FFD: 251 vs. 257 ms, b = -0.03, SE = 0.02, $t = -1.74$; SFD: 250 vs. 257 ms, $b = -0.03$, SE = 0.02, $t = -1.83$) and significant for GD and TVT (GD: 272 vs. 289 ms, $b = -0.06$, SE = 0.02, $t = -3.39$; TVT: 302 vs. 334 ms, $b = -0.08$, SE = 0.02, $t = -3.44$). The predictability effect (predictable vs. unpredictable) was significant for all types of duration measures (FFD: 249 vs. 259 ms, $b = -0.04$, SE = 0.01, $t = -3.72$; SFD: 248 vs. 260 ms, $b = -0.05$, SE = 0.01, $t = -4.15$; GD: 272 vs. 288 ms, $b = -0.06$, SE = 0.01, $t = -4.32$; TVT: 307 vs. 329 ms, $b = -0.07$, SE = 0.02, $t = -4.45$). There was no interaction effect of word frequency and word predictability for any of the duration measures.

Target Fixation Probabilities. The analysis of target fixation probabilities was conducted based on 3172 and 2470 observations for the skipping rate and refixation rate, respectively. Generally, high-frequency or predictable words exhibited a higher skipping rate and a lower refixation rate compared to low-frequency or unpredictable words. The word frequency effect (high frequency vs. low frequency) was significant

Table 5.2 Mean fixation durations and probabilities (standard error) on target words for each condition

	High frequency		Low frequency	
	Predictable	Unpredictable	Predictable	Unpredictable
		Fixation Duration (ms)		
SFD	245 (3.72)	256 (3.69)	250 (3.74)	264 (4.08)
FFD	247 (3.60)	254 (3.40)	251 (3.50)	263 (3.60)
GD	266 (4.64)	276 (4.45)	278 (4.59)	300 (4.90)
TVT	294 (6.18)	308 (5.79)	319 (7.19)	348 (7.30)
		Fixation Probability (%)		
Skipping rate	0.25 (0.02)	0.22 (0.01)	0.22 (0.01)	0.18 (0.01)
Refixation rate	0.09 (0.01)	0.10 (0.01)	0.14 (0.01)	0.18 (0.02)

Note FFD = first fixation duration; SFD = single fixation duration; GD = Gaze Duration; TVT = total viewing time

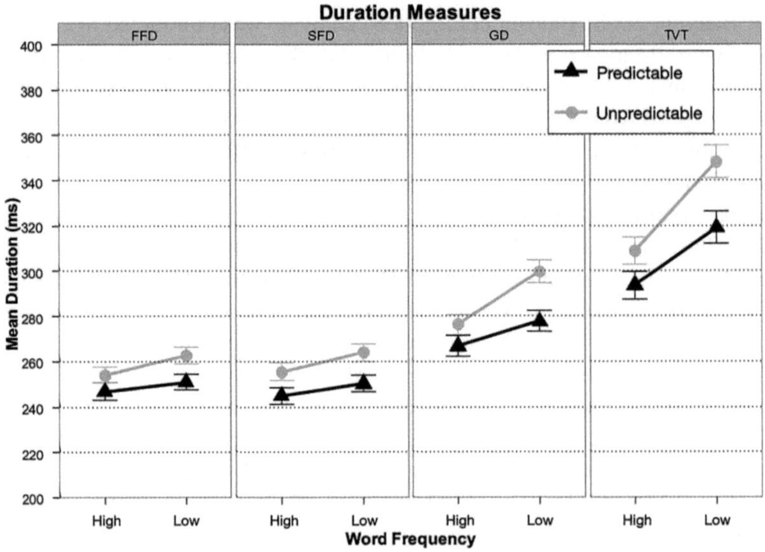

Fig. 5.1 Mean fixation durations of the first fixation duration (FFD), single fixation duration (SFD), gaze duration (GD), and total viewing time (TVT) as a function of word frequency and word predictability

for both skipping and refixation rates [Skipping: 24% vs. 20%, $b = 0.23$, SE = 0.1, $z = 2.26$, $p < 0.05$; Refixation: 10% vs. 16%, $b = -0.65$, SE = 0.13, $z = -4.9$, $p < 0.01$]. The predictability effect (predictable vs. unpredictable) was significant for the skipping rate [24% vs. 20%, $b = 0.21$, SE = 0.09, $z = 2.40$, $p < 0.05$] but insignificant for the refixation rate [12% vs. 14%, $b = -0.2$, SE = 0.13, $z = -1.52$, $p = 0.13$]. None of the interaction effects were significant for probability measures.

5.4 Discussion

The study demonstrated both word frequency and word predictability effects on first-pass durations (FFD, SFD, and GD) and total viewing time (TVT). Chinese readers allocated less time to high-frequency or predictable words compared to low-frequency or unpredictable words. Regarding the skipping rate, both predictability and frequency effects were significant, with similar effect sizes (3% vs. 4%). Notably, significant word predictability effects were observed for first-pass FFD and SFD, marking the first instance of such effects in Chinese reading research. This finding is corroborated by an ERP study utilizing a similar design and materials, which also revealed predictability effects on early stage components such as the anterior N1 and P200 (Lee et al., 2012). Additionally, the word predictability effects on fixation durations and the skipping rate align with findings from English studies

(Miellet et al., 2007; Rayner et al., 2004). Overall, this consistent evidence suggests that Chinese readers rely on contextual information for word recognition prior to integrating meaning for comprehension.

Nonetheless, the word predictability effect exhibited greater reliability compared to the word frequency effect. The significant predictability effect was approximately double the magnitude of the less consistent frequency effect (FFD: 10 ms vs. 6 ms; SFD: 12 ms vs. 7 ms, respectively). This suggests that Chinese readers may rely more heavily on contextual information than readers of alphabetic languages. Several unique characteristics of the Chinese writing system contribute to this reliance. For instance, the absence of spaces between words in Chinese sentences creates ambiguity regarding word boundaries. Additionally, the majority of Chinese words are compounds, making it challenging to discern the relationship between a word and its constituent characters in terms of both orthographical and morphological structures. These perceptual and lexical ambiguities must be resolved to accurately extract words from the character string. Inhoff and Wu (2005) conducted an eye movement study of reading Chinese, manipulating a four-character sequence (C1234) where C12 and C34 formed context-consistent words in a sentence. They found that readers spent more time processing the four-character sequence when the two center characters (C23) formed a legal word in the ambiguous condition. Consequently, determining the location of Chinese words within a sentence is not as straightforward as in alphabetic languages with word spaces. Chinese readers must effectively utilize contextual information to efficiently identify word boundaries and extract the correct word information.

The skipping rate exhibited both predictability and frequency effects, consistent with findings from previous studies (Rayner & Well, 1996; Rayner et al., 2004, 2005). This suggests that contextual information plays a role in the early stages of reading, particularly in parafoveal preview, influencing whether the target word is skipped. It's plausible that the word is identified parafoveally and subsequently skipped due to facilitated lexical access (Rayner et al., 2001) or a mechanism of hypothesis generation based on incomplete information (Brysbaert & Vitu, 1998). Alternatively, this effect may involve the retrieval of word meaning from memory based on context, without accessing parafoveal information (Kliegl et al., 2006).

However, these results regarding the skipping rate are inconsistent with those of Rayner et al. (2004), who reported an interaction effect, and Miellet et al. (2007), who found no effect of frequency and predictability on the skipping rate. Such discrepancies across studies could stem from variations in sentence difficulty. Difficult sentences may lower the skipping rate and hinder potential word frequency or predictability effects (Miellet et al., 2007). Further investigation is needed to determine whether the inconsistencies between Chinese and English studies are attributable to the strong reliance on context to resolve Chinese word boundary ambiguity.

The present data across all eye movement measures showed no interaction between word predictability and word frequency. This absence of interaction aligns with findings from previous English studies, both involving corpus data (Calvo & Meseguer, 2002; Kliegl et al., 2004, 2006) and experimental data via direct manipulation

(Rayner et al., 2006). The additive effects observed suggest that the influence of both factors on lexical access is either mutually independent or occurs at different stages (Staub, 2015). Interestingly, an ERP study of reading Chinese noted an interaction effect between word predictability and word frequency, specifically on the anterior N1 component, although not on the P200 or N400 components (Lee et al., 2012). This early interaction effect, occurring within 100–200 ms after perceiving a word, suggests that contextual information may directly facilitate word processing at the perceptual level. Furthermore, a recent study investigated these effects on developing Chinese readers and found interaction effects on skipping rates and reading times (Liu et al., 2021). Notably, they observed predictability effects on eye movement measures for low-frequency words but not for high-frequency words. One possible explanation for the observed interaction effect among developing readers, but not among adult readers, could be attributed to individual differences in reading ability. For developing readers, the efficiency of lexical access may require more direct support from contextual information compared to adult readers. In contrast, for adult readers, contextual information interacts with word familiarity during perceptual analysis while processing a blurred or partially presented word in the parafovea.

5.5 Conclusion

In summary, contextual information is crucial in reading Chinese sentences, aiding both lexical access for word recognition and meaning integration for comprehension. Evidence from word skipping and other measures demonstrates that context influences both perceptual and lexical processing during the reading of Chinese sentences. Further research is needed to directly investigate when and how contextual predictability resolves word ambiguities in reading Chinese sentences, particularly in both foveal and parafoveal processing. This would provide deeper insights into the mechanisms underlying Chinese reading comprehension.

Acknowledgements This work was supported by grants from Taiwan's National Science and Technology Council (NSC 99-2410-H-004-091-MY2 and NSTC 110-2410-H-004 -112-MY2). The data presented in this work has been previously disclosed in the project report of grant NSC 99-2410-H-004-091.

References

Academia Sinica Balanced Corpus. (2004). [CDROM]. Academia Sinica, Taipei, Taiwan.

Altarriba, J., Kroll, J., Sholl, A., & Rayner, K. (1996). The influence of lexical and conceptual constraints on reading mixed-language sentences: Evidence from eye fixations and naming times. *Memory and Cognition, 24*, 477–492.

Baayen, R. H. (2008). *Analyzing linguistic data: A practical introduction to statistics using R*. Cambridge University Press.

Baayen, R. H., Davidson, D. J., & Bates, D. M. (2008). Mixed-effects modeling with crossed random effects for subjects and items [Article]. *Journal of Memory and Language, 59*(4), 390–412. https://doi.org/10.1016/j.jml.2007.12.005

Balota, D. A., Pollatsek, A., & Rayner, K. (1985). The interaction of contextual constraints and parafoveal visual information in reading. *Cognitive Psychololgy, 17*(3), 364–390. https://doi.org/10.1016/0010-0285(85)90013-1

Bates, D. M., & Maechler, M. (2010). *lme4: Linear mixed-effects models using S4 classes.*

Brysbaert, M., & Vitu, F. (1998). Word skipping: Implications for theories of eye movement control in reading. In G. Underwood (Ed.), *Eye guidance in reading and scene perception* (pp. 125–147). Elsevier Science Ltd.

Calvo, M. G., & Meseguer, E. (2002). Eye movements and processing stages in reading: Relative contribution of visual, lexical and contextual factors. *The Spanish Journal of Psychology, 5*(1), 66–77.

Ehrlich, S. F., & Rayner, K. (1981). Contextual effects on word perception and eye movements during reading. *Journal of Verbal Learning and Verbal Behavior, 20*(6), 641–655.

Engbert, R., Nuthmann, A., Richter, E. M., & Kliegl, R. (2005). SWIFT: A dynamical model of saccade generation during reading. *Psychological Review, 112*(4), 777–813. https://doi.org/10.1037/0033-295x.112.4.777

Gollan, T. H., Slattery, T. J., Goldenberg, D., Van Assche, E., Duyck, W., & Rayner, K. (2011). Frequency drives lexical access in reading but not in speaking: The frequency-lag hypothesis. *Journal of Experimental Psychology: General, 140*(2), 186–209. https://doi.org/10.1037/a0022256

Hand, C. J., Miellet, S., O'Donnell, P. J., & Sereno, S. C. (2010). The frequency-predictability interaction in reading: It depends where you're coming from. *Journal of Experimental Psychology: Human Perception and Performance, 36*(5), 1294–1313. https://doi.org/10.1037/a0020363

Inhoff, A. W. (1984). Two stages of word processing during eye fixations in the reading of prose. *Journal of Verbal Learning and Verbal Behavior, 23*(5), 612–624.

Inhoff, A. W., & Wu, C. (2005). Eye movements and the identification of spatially ambiguous words during Chinese sentence reading. *Memory and Cognition, 33*(8), 1345–1356.

Kliegl, R., Grabner, E., Rolfs, M., & Engbert, R. (2004). Length, frequency, and predictability effects of words on eye movements in reading. *European Journal of Cognitive Psychology, 16*(1–2), 262–284. https://doi.org/10.1080/09541440340000213

Kliegl, R., Nuthmann, A., & Engbert, R. (2006). Tracking the mind during reading: The influence of past, present, and future words on fixation durations. *Journal of Experimental Psychology: General, 135*(1), 12 35. https://doi.org/10.1037/0096-3445.135.1.12

Lee, C. Y., Liu, Y. N., & Tsai, J. L. (2012). The time course of contextual effects on visual word recognition. *Frontiers in Psychology, 3.* https://doi.org/10.3389/fpsyg.2012.00285

Liu, N. N., Wang, X., Yan, G. L., Paterson, K. B., & Pagán, A. (2021). Eye movements of developing chinese readers: Effects of word frequency and predictability. *Scientific Studies of Reading, 25*(3), 234–250. https://doi.org/10.1080/10888438.2020.1759074

Miellet, S., Sparrow, L., & Sereno, S. C. (2007). Word frequency and predictability effects in reading French: An evaluation of the EZ Reader model. *Psychonomic Bulletin and Review, 14*(4), 762.

R Development Core Team. (2010). *R: A language-and environment for statistical computing.* In [Computer software]. R Foundation for Statistical Computing.

Rayner, K., Ashby, J., Pollatsek, A., & Reichle, E. D. (2004). The effects of frequency and predictability on eye fixations in reading: Implications for the EZ Reader model. *Journal of Experimental Psychology: Human Perception and Performance, 30*(4), 720–732.

Rayner, K., Binder, K. S., Ashby, J., & Pollatsek, A. (2001). Eye movement control in reading: Word predictability has little influence on initial landing positions in words. *Vision Research, 41*(7), 943–954.

Rayner, K., Li, Z., Juhasz, B. J., & Yan, G. (2005). The effect of word predictability on the eye movements of Chinese readers. *Psychonomic Bulletin and Review, 12*(6), 1089.

Rayner, K., Reichle, E. D., Stroud, M. J., Williams, C. C., & Pollatsek, A. (2006). The effect of word frequency, word predictability, and font difficulty on the eye movements of young and older readers. *Psychology and Aging, 21*(3), 448–465. https://doi.org/10.1037/0882-7974.21.3.448

Rayner, K., & Well, A. D. (1996). Effects of contextual constraint on eye movements in reading: A further examination. *Psychonomic Bulletin and Review, 3*(4), 504–509.

Reichle, E. D., Rayner, K., & Pollatsek, A. (2003). The E-Z reader model of eye-movement control in reading: comparisons to other models. *Behavioral and Brain Sciences, 26*(4), 445–476; discussion 477–526.

Stanovich, K. E., & West, R. F. (1983). On priming by a sentence context. *Journal of Experimental Psychology: General, 112*(1), 1–36.

Staub, A. (2015). The Effect of lexical predictability on eye movements in reading: Critical review and theoretical interpretation. *Language and Linguistics Compass, 9*(8), 311–327. https://doi.org/10.1111/lnc3.12151

Tsai, J.-L. (2014). Eye movement guidance in reading unspaced text in Thai and Chinese. In H. Winskel & P. Padakannaya (Eds.), *South and Southeast Asian psycholinguistics* (pp. 265–271). Cambridge University Press.

Tsai, J.-L., Lee, C. Y., Yen, M. H., Chan, W. H., & Yang, Y. J. (2009). *Effects of word predictability and word frequency on eye movements in reading Chinese sentences.* The 13th International Conference on the Processing of East Asian Languages (ICPEAL), Beijing, China.

Tsai, J. L., Lee, C. Y., Lin, Y. C., Tzeng, O. J. L., & Hung, D. L. (2006). Neighborhood size effects of Chinese words in lexical decision and reading. *Language and Linguistics, 7*, 659–675.

Tsai, J. L., & McConkie, G. W. (2003). Where do Chinese readers send their eyes. In *The mind's eye: Cognitive and applied aspects of eye movement research* (pp. 159–176).

Yan, G., Tian, H., Bai, X., & Rayner, K. (2006). The effect of word and character frequency on the eye movements of Chinese readers. *British Journal of Psychology, 97*, 259–268.

Yang, H. M., & McConkie, G. W. (1999). Reading Chinese: Some basic eye-movement characteristics. In *Reading Chinese script: A cognitive analysis* (pp. 207–222).

Yen, M. H., Radach, R., Tzeng, O. J. L., Hung, D. L., & Tsai, J. L. (2009). Early parafoveal processing in reading Chinese sentences. *Acta Psychologica, 131*(1), 24–33. https://doi.org/10.1016/j.actpsy.2009.02.005

Chapter 6
The Foreign Language Effect: Beyond Language

Angela K. Tzeng and Shih-Chieh Lai

Abstract Are you willing to sacrifice one person to save five people? In this classical moral dilemma, bilinguals tend to make different decisions in different languages. The Foreign Language effect (FLe) refers to the phenomenon of bilinguals processing the same information differently when using their native language (NL) and foreign language (FL). This effect was repeatedly found in moral dilemma judgment and decision tasks using questions like the Asian Disease Problem or the Footbridge Dilemma. While using NL produces more deontological choices, using FL produces more utilitarian decisions. One common explanation of the FLe is that NL involves more emotional reactions. Another common explanation is that people tend to use System 1 while making decisions with NL. In addition to emotion and thinking processes, recent studies show that accessibility of information may also be important in FLe. Thirty-nine Chinese-Japanese-English trilinguals were recruited in the current study. The procedure in use is a Semantic Matching Judgment Task. There will be a word presented visually and auditorily simultaneously followed by a picture in each trial. Participants needed to decide whether or not the word and the picture matched semantically. Accessibility will be investigated by language proficiency, semantically matched or not in stimuli, and cultural congruence that goes beyond language. We found that both language proficiency and semantical matching had an impact on FLe in Experiment 1. Cultural Congruency is manipulated by choosing conflicting pictures from Taiwan and Japan. Congruency and language in use showed significant interaction in Experiment 2. This suggests that context produces an FLe effect that works beyond language itself. In sum, the FL proficiency of the trilinguals in the current study is equivalent, but they are familiar with Japanese culture than with US culture. In other words, different patterns in their performance ought to come from factors beyond language per se. Our result supports the notion of accessibility on FLe.

A. K. Tzeng (✉) · S.-C. Lai
Department of Psychology, Chung Yuan Christian University, Taoyuan, Taiwan
e-mail: angelat84@gmail.com

A. K. Tzeng
Center for Theoretical Science, Chung Yuan Christian University, Taoyuan, Taiwan

D. H. Wu (ed.), *Cognitive and Neural Foundations of Chinese Reading*, Chinese Language Learning Sciences, https://doi.org/10.1007/978-981-96-6669-0_6

Keywords Foreign language effect · Trilingualism · Contextual effect · Semantic judgment

6.1 Introduction

6.1.1 Foreign Language Effect (FLe)

Foreign language effect (FLe) refers to the phenomenon of bilinguals processing the same information differently in their native language (NL) or foreign language (FL). FLe was first reported by Keysar et al. (2012) using the Asian Disease Problem adopted from Tversky and Kahneman (1981). They asked English–Spanish bilinguals to complete six decision-making experiments, including the Asian Disease Problem, Loss Aversion, Myopic Loss Aversion, and Consequential Investment Behavior. The main finding was that the Framing Effect and Loss Aversion were significantly reduced in the FL condition. FLe were then found in the areas such as moral judgment, moral decision, as well as risk perception (Costa et al., 2014a, 2014b; Geipel et al., 2015a, 2015b, 2016; Keysar et al., 2012).

In classical moral dilemmas like the Footbridge Problem, bilinguals tend to produce more deontological choices in the NL condition and more utilitarian decisions in the FL condition (Costa et al., 2014b). A utilitarian choice in the Footbridge Problem means the participants choose to physically push a stranger down from the footbridge to save five strangers. This makes the Footbridge Problem more emotionally charged than the original Trolley Problem. Therefore, emotional involvement has been proposed as one of the common explanations for FLe.

Why would emotional involvement affect people's decisions when they use different languages facing the same problem? Gao et al. (2015) investigated the "hot hand fallacy" of bilinguals. The hot hand fallacy refers to the positive feedback bias of people to overestimate the probability of winning in the next hand. The result showed more hot hand fallacy in the NL condition. They suggested the results mean less emotional bias in the FL condition.

FLe is often about making decisions. Will people think differently in the NL and FL conditions? Kahneman (2003) proposed the dual-process theory and stated two thinking systems. System 1 is more intuitive, fast, automatic, effortless, and affect-driven, therefore making it harder to intrude during the process. On the other hand, System 2 involves more reasoning processing. It is slower, controlled, effortful, and bound by an individual's awareness of decision and attitude. Based on this distinction, it is reasonable to argue that System 1 is more dominant in the NL condition because people are more familiar with and proficient in, and also more emotionally charged in NL (Geipel et al., 2015a; Li et al., 2014; Malt et al., 2015). On the contrary, System 2 may be more dominant in the FL condition because the usage of FL creates more psychological distance, less emotional reaction, and allows more space for utilitarian consideration. The dual-process theory then became another common account for FLe (Cipolletti et al., 2016).

The researchers adopting the dual-process model for the FLe took further steps for underlying mechanisms. Hayakawa et al. (2017) argued that more utilitarian choices occur in FL not because people think deeper and more cautiously, as in System 2. It happens because of the reduction of aversive emotion in the FL condition. In other words, the usage of FL eases off the resistance to maximize the utilitarian outcome. Muda et al. (2018) argued that FLe reflects the reduction of emotion in the FL condition. This reduction of emotion allows participants not to consider all potential harm, thus more utilitarian choices. Corey et al. (2017) carried out a series of nine experiments and concluded that FLe is not due to the increase in cognitive control but due to the reduction of emotion and psychological distance.

Neurological evidence was also reported to examine dual processes in FLe. Greene (2014) proposed that utilitarian choice is mainly an automatic emotional reaction, whereas deontological choice is made by conscious logic and cognitive control. In an fMRI study, Greene et al. (2001) found personal dilemmas induced more activity in the amygdala, posterior cingulate gyrus, and angular gyrus when contrasted with impersonal and non-moral dilemmas. These areas are related to emotional reactions. On the contrary, brain areas that are more related to working memory, such as the middle frontal gyrus, showed less activation in personal dilemmas. Greene et al. (2004) also found stronger activation in the anterior cingulate cortex (ACC, for processing cognitive conflict) and dorsolateral prefrontal cortex (DLPFC, for abstract reasoning) while making deontological choices.

6.1.2 Inconsistency in the FLe

As the literature grows, there is some inconsistency in the FLe. For example, does emotional involvement matter? Many studies show that FLe is related to emotional involvement (Caldwell-Harris, 2015; Caldwell-Harris & Ayçiçeği-Dinn, 2009; Hadjichristidis et al., 2015; Wu & Thierry, 2010, 2012), but some don't. Costa et al. (2014b) found no FLe in the Trolley Problem, yet a significant effect in the Footbridge Problem with English–Spanish and Spanish–English bilinguals. Geipel et al. (2015b) manipulated emotional involvement while trying to replicate Costa et al. (2014b). Among two personal and two impersonal dilemmas, only one in each category showed FLe. In other words, the role of emotion plays is inconsistent.

There is also an inconsistent finding of FLe in dual-process theory. Keysar et al. (2012) suggested that decisions made with NL are automatic, faster, emotional, and intuitive. The decisions with FL, however, are slower, more logical, and unbiased. Yet, Costa et al. (2014b) didn't find consistent results in various heuristic biases. Geipel et al. (2015b) asked participants to also rate their confidence while judging the moral wrongness of several private actions, such as consensual insect. The results showed less severe moral judgments and lower confidence in the FL condition, which contradicted the notion that FL facilitated deliberation. Geipel et al. (2015b) therefore proposed a "reduced intuition account" to explain the usage of FL makes people less averse toward the behavior in question, and the evaluation of the intention was also reduced. They concluded that the FLe is best explained by a reduced activation of social and moral norms. Hayakawa et al. (2017) also showed that the usage of FL does not increase utilitarianism. Instead, FL increases blunted deontology.

6.1.3 Accessibility

Beyond the emotional and dual-process model, there are other perspectives to look at the FLe. Accessibility of information is a more recent account. The role accessibility plays in FLe (Chan et al., 2016; Geipel et al., 2015a, 2015b; Miozzo et al., 2020) has been investigated by various research groups.

Miozzo et al. (2020) directly address the role of emotion by conducting four experiments with Italian-Venetian and Italian-Bergamasque balanced bilinguals. They successfully reproduced FLe in Venetian and Bergamasque conditions with the Asian Disease Problem and the Footbridge Problem. They emphasized that this finding does not support the emotional account because the participants showed equal emotional intensities in their NL (Italian) and FLs (Venetian or Bergamasque). The difference between the NL and FL is the cultural context. For these participants, Venetian or Bergamasque are used in informal household situations, whereas Italian is used in a more formal context. NL is therefore closely connected with social and moral norms. In sum, the FLe can be accounted for by the accessibility of related information.

Other researchers (Knez & Nordhall, 2017; Körner et al., 2016; Kusev et al., 2012) also point out that the role emotion plays may very much come from memory and experience. After all, memory includes the context of how people acquire and use their languages.

To emphasize that language and memory are not independent of each other, Marian and Neisser (2000) conducted a language-dependent recall study. They proposed that language can create Mental Reinstatement of Context (MRC), which will make some memories more accessible than others (Marian & Neisser, 2000; Matsumoto & Stanny, 2006; Mortensen et al., 2015).

Marian and Neisser's (2000) idea is: (1) when people use a language, the words in use work like physical stimuli which can create external context; (2) their mental processes are related to internal context; and (3) when trying to memory certain events, using the same language now (retrieval) and then (encoding) can create MRC which makes the memory more accessible than others. They asked Russian-English immigrants to recall autobiographical memories in Russian and English. The result demonstrated a significant language-dependent effect. This language-dependent effect is also supported by Japanese-English non-immigrants study (Matsumoto & Stanny, 2006).

Ellis et al. (2018) recruited Welsh-English bilinguals to investigate how context affects their decision. They presented sentences related to Welsh culture or not in both English and Welsh. Participants were to judge the sentences to be true or false. The result showed that participants' judgments were more accurate in the NL condition (Welsh), whether the sentences were presented affirmatively or negatively. When the sentences were presented in English, only affirmative ones were accurately judged. It seems that FLe not only appears in decision but also perception of factual information.

In sum, to make a decision, a person first needs to retrieve related information from memory. Since it is quite common for a person's NL and FL to be acquired and used in different contexts, different decisions may be made simply because of the choice of language. Thus, FLe. Generally speaking, NL is used in more formal situations and is connected with social norms. This may be why more deontological choices are found in the NL condition.

To further investigate the role of accessibility on FLe, Chinese-Japanese-English trilinguals were recruited in the current study. A critical point of the current study is the unique relationship between Taiwanese and Japanese cultures. These two cultures are closely related due to their geographical and historical connectedness. For example, there is the "lion dance" (舞獅 in Chinese) in festivals, and the "rice cake" (年糕 in Chinese) and the "red envelope" (紅包 in Chinese) for New Year in both cultures. Yet, these shared cultures are acquired and performed in different contexts. That made these items suitable for seeing how accessibility influences people's decisions.

All participants in the current study are fluent in both FLs (Japanese and English) but more familiar with Japanese culture than US culture. The difference we might find between Japanese and English conditions can be explained by accessibility (culture) rather than FLs per se.

There were two experiments in the current study. The Semantic Matching Judgment Task was used in both experiments. A word was presented visually and auditorily in each trial, followed by a picture. Their task was to judge whether the word and the picture were semantically matched. In Experiment 1, participants completed this task in three language blocks. Only Taiwanese and Japanese stimuli were used in Experiment 2 because the key variable here is the congruency between the words and the pictures. The word and the picture are congruent if they are from the same culture (more details in the methods section). The accessibility should be higher in the congruent condition. The role of accessibility on FLe was then established.

6.2 Methods

6.2.1 Stimuli Selection

33 college students (14 males and 19 females) were recruited from Chung-Yuan Christian University for stimuli selection. They were born in Taiwan with Mandarin Chinese as their NL. Upon completing the task, they received compensation for their work.

240 pictures from categories of food, clothing, religions, history, architecture, and festivals were selected from open-source archive or pay-for-permission websites: Shutterstock, Pixabay, Pakutaso, and photo-AC. There were 90 pictures related to Taiwanese culture, 90 pictures related to Japanese culture, and 60 pictures related to US culture. Participants were asked three questions on a 7-point scale:

Q1: How much can this photo represent Taiwanese (Japanese or US) culture?
Q2: How much do you think this photo also fits in other cultures?
Q3: Is "xxx" (object name in Chinese) suitable for this photo?

To be selected for our experiments, all photos must score high on Q1 for representativeness and low on Q2 for distinctiveness. The task took 30–50 min to complete. Ninety photos were chosen for Experiment 1 (30 in each language) and 62 photos for Experiment 2 (31 from Chinese and 31 from Japanese culture).

6.3 Experiment 1: Semantic Matching Judgment Task in Three Languages

6.3.1 Participants

Thirty-nine Chinese-Japanese-English trilinguals (14 males and 25 females) who were not recruited for the stimuli selection procedure volunteered in this experiment. Their ages are 22 and 44 (mean = 30.56, SD = 60.06). Mandarin-Chinese is the NL for all participants. They would receive compensation for participation.

6.3.2 FL Proficiency and Cultural Familiarity

For Japanese, all participants have a vocabulary size of over 20,000 words (M = 28,854.46, SD = 5637.61), equivalent to Japanese Adults. Eight participants (20.5%) pass JLPT (Japanese-Language Proficiency Test) N2 (equivalent to CEFR B2); Thirty (76.9%) pass JLPT N1 (equivalent to CEFT C2). They also have composite factor scores in LSBQ higher than −3.13 indicating they are balanced or receptive bilinguals. For Chinese-Japanese LSBQ, there are 28 balanced bilinguals (71.8%) and 11 receptive bilinguals (28.2%).

For English, 13 participants (33.3%) haven't obtained formal certificates. The other 26 participants passed various CEFR-equivalent tests: 3 (7.7%) have A2, 11 (28.2%) have B1, 5 (12.8%) have B2, 6 (15.4%) have C1, and 1 (2.6%) has the highest C2 certificate.

All participants were asked to evaluate their cultural familiarity (see Table 6.1).

Table 6.1 Self-report cultural familiarity of Chinese-Japanese-English Trilinguals ($N = 39$)

	Familiarity				
Cultures	Very unfamiliar	Unfamiliar	Average	Familiar	Very familiar
Taiwan	0	0	3 (7.7%)	14 (35.9%)	22 (56.4%)
Japan	0	1 (2.6%)	3 (7.7%)	31 (79.5%)	4 (10.3%)
US	6 (15.4%)	15 (38.5%)	11 (28.2%)	7 (17.9%)	0

6.3.3 Stimuli

Ninety pictures (30 in each language) were selected. A corresponding word in Chinese was determined from the previous stimuli selection procedure.

6.3.4 Design and Procedure

A 3 (Language) × 2 (Semantic Matching) block design was used in Experiment 1. There were 60 trials in each block. The pictures and the words were semantically matched in half of the trials. All trials were presented randomly within the block. The order of the languages was counterbalanced. Each trial started with a fixation point (500 ms), followed by a word presented visually and auditorily at the same time (1000 ms), and a picture (3000 ms, or a response was made). Participants were to decide whether the word and the picture were semantically matched. Figure 6.1 shows the procedure for two examples in the Taiwanese block. "Lion Dance" (舞獅) matches the picture, whereas "Mask" (面具) does not.

6.4 Results

6.4.1 Reaction Time

RTs in Chinese are 687 ms (SD = 141 ms) for the Matched condition and 708 ms (SD = 139 ms) for the Unmatched condition, respectively. In the Japanese block, 810 ms (SD = 196 ms) for the Matched condition and 798 ms (SD = 160 ms) for the Unmatched condition; in the English block, 838 ms (SD = 178 ms) for the Matched condition and 776 ms (SD = 148 ms) for the Unmatched condition (see Fig. 6.2).

A 2 × 3 ANOVA shows a significant main effect of Language $F\ (2, 76) = 10.48$, $P < 0.001$, $\eta_{p^2} = 0.216$. Scheffé's test indicates RT in Chinese is the fastest, whereas no significant difference in RT between Japanese and English conditions. Semantic Matching has no main effect. There was an interaction of Language × Matching $F(2, 76) = 14.27, P < 0.001, \eta_{p^2} = 0.273$.

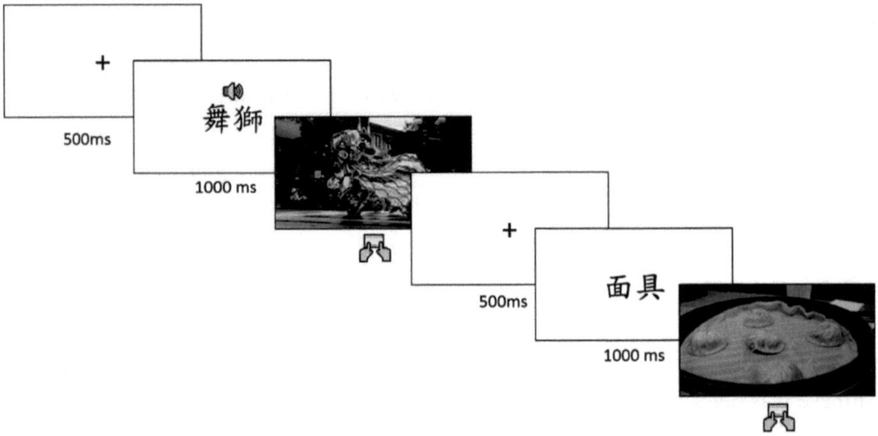

Fig. 6.1 The procedure of the semantic matching judgment task in experiment 1 in the Chinese block. *Note.* These are two examples in this figure. Each trial contains 3 frames starting with a fixation (500 ms), a word (1000 ms), and a picture (up to 3000 ms). The answer is "yes" for the first trial and "no" for the second trial

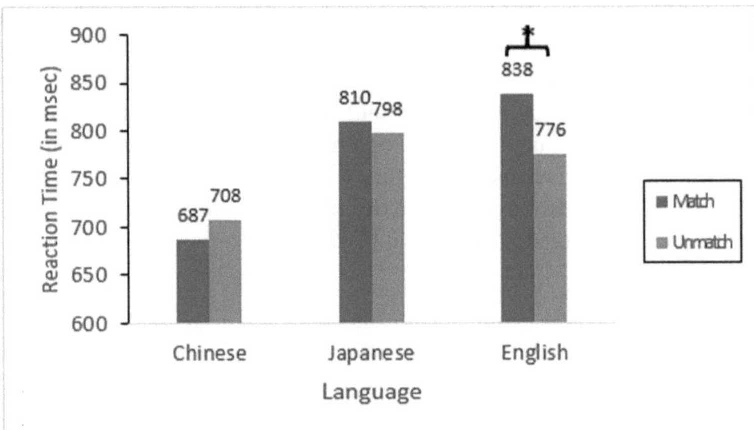

Fig. 6.2 Reaction Time (in msec) for semantic matching judgment task in experiment 1. *Note* The RTs in three languages are used to judge whether the words and the pictures are semantically matched

Language shows a significant, simple main effect in the Matched condition $F(2, 152) = 16.16, P < 0.001, \eta_{p^2} = 0.175$; and in Unmatched condition $F(2, 152) = 5.39, P = 0.005, \eta_{p^2} = 0.066$. For both conditions, Chinese RT was the fastest among the three languages, and there was no difference between Japanese and English. There was also a simple main effect in English between Matched and Unmatched conditions, $F(1, 114) = 21.27, P < 0.001, \eta_{p^2} = 0.157$. The matched condition was slower than the Unmatched condition. No effect was found between the Matched and the Unmatched conditions in Japanese and Chinese.

6.4.2 *Accuracy*

Accuracy in Chinese is 98.6% (SD = 2.2%) for the Matched condition and 98.6% (SD = 2.1%) for the Unmatched condition, respectively; in Japanese, 96.8% (SD = 4.4%) for the Matched condition and 97.4% (SD = 3.4%) for the Unmatched condition; and in English, 93.0% (SD = 5.2%) for the Matched condition and 97.0% (SD = 3.1%) for the Unmatched condition (see Fig. 6.3).

A 2 × 3 ANOVA shows a significant main effect of Language $F(2, 76) = 16.71$, $P < 0.001$, $\eta_{p^2} = 0.305$. Scheffé's test indicates Chinese has the highest accuracy rate (98.6%), followed by Japanese (97.1%), the English (95.0%) yields the lowest accuracy rate. There is also a significant main effect of Semantic Match $F(1, 38) = 11.90, P = 0.001, \eta_{p^2} = 0.238$, as well as a significant interaction between Language and Semantic Matching $F(2, 76) = 9.60, p = 0.001, \eta_{p^2} = 0.202$.

Language shows a significant, simple main effect in the Matched condition $F(1, 114) = 21.02, P < 0.001, \eta_{p^2} = 0.249$. with Chinese (98.6%) higher than Japanese (96.8%), and higher than English (93.0%). No simple main effect in the Unmatched condition.

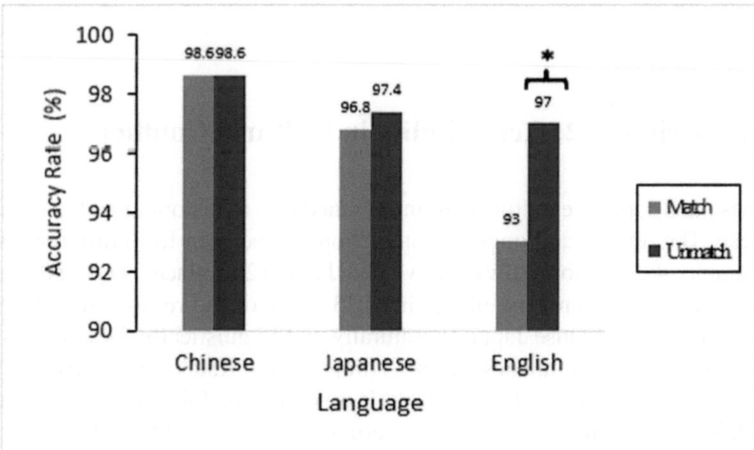

Fig. 6.3 Accuracy rate (%) for semantic matching judgment task in experiment 1. *Note* The accuracy in three languages to judge whether the words and the pictures are semantically matched

6.4.3 Discussion

The primary purpose of Experiment 1 is to establish the relationship between language proficiency and information accessibility.

Unsurprisingly, participants performed the best on RT and accuracy in the NL (Chinese) block. All participants were fluent in Japanese and English because they all passed various tests to equivalent standards, and Scheffé's test showed no difference in their RT.

The role accessibility plays can be established by comparing the patterns between Japanese and English blocks. In the English block, the Matched condition yielded the slowest RT (838 ms) and the lowest accuracy rate (93%). In other words, participants performed well on the Unmatched English conditions (correct rejection) but not so well on the Match condition (hits). Since their English is proficient, the relatively poor performance on the Matched condition could be attributed to the accessibility. The lowest familiarity with US culture from Table 6.1 supports this notion, too. All 90 pictures and the corresponding words are from one of the three cultures. Participants only needed to judge a particular picture or a word was from Taiwanese or Japanese cultures to make a correct rejection. On the other hand, a hit required accessing accurate information. The low familiarity with US culture made it most difficult for the participants.

In sum, all participants were fluent in Japanese and English. However, they were familiar with Japanese culture rather than with US culture. The discrepancy in performance between the two FLs allows us to tease apart the role that language proficiency and accessibility play. In Experiment 2, we further examined the accessibility under cultural conflict situations.

6.5 Experiment 2: Accessibility in Cultural Conflict

Taiwanese and Japanese cultures are intertwined due to historical and geographical closeness. The unique and ongoing interaction can be seen from different aspects. For example, 4.2 million Taiwanese visited Japan (2nd place), and 580 thousand Japanese visited Taiwan (2nd place) in 2023—one of the reasons that Taiwanese love to visit Japan because Japan is culturally and linguistically friendly. There are three writing systems in Japanese: Kanji, Hiragana, and Katakana. Kanji and Chinese characters are closely related. This makes it very easy for Taiwanese people to travel to Japan and understand the culture to a certain degree, even if they do not speak the language.

The accessibility issue is more interesting because the shared cultures are also very context-dependent. For example, it is a custom for Taiwanese and Japanese to give their family an envelope with money in the New Year. It is called the "red envelope" because it has to be red and the amount of cash inside has to be in even number. However, the envelope has to be white and the amount of money has to

Table 6.2 An example of pictures in cultural congruent or incongruent condition in experiment 2: New Year money

	Culture	
Context	Taiwan	Japan
Congruent	 壓歳銭 (Chinese)	 おとしだま (Japanese)
Incongruent	 おとしだま (Japanese)	 壓歳銭 (Chinese)

be an odd number (5 being the best) as in Table 6.2. Both cultures have New Year envelopes, so it is easy for Taiwanese and Japanese to understand the meaning, but it would be very offensive if it is carried out wrongly (e.g., wrong colors).

This shared but not identical culture provides an excellent opportunity to see how accessibility works in FLe.

6.5.1 *Participants*

The same 39 participants were invited back two days after they completed the first experiment.

6.5.2 *Stimuli*

Thirty-one pairs of pictures were chosen. In each pair, one picture represents Taiwanese culture and the other represents Japanese culture. The two pictures are about the same concept but conflict in context.

6.5.3 Design and Procedure

A 2 (Language) × 2 (Cultural Congruency) × 2 (Semantic Matched) block design was used. Cultural Congruency was defined as responding to a culture-dependent picture in its corresponding language (e.g., Deciding on the Chinese New Year Envelope in Chinese). Half of the participants did the Chinese block first; the other half did the Japanese block first. All items were presented randomly within the block. The procedure started with a fixation point (500 ms) followed by a word presented both visually and auditorily (1000 ms), a picture then appeared (3000 ms or the response was made). Participants needed to perform a Semantic Matching Judgment Task by pressing key J for "yes" or key F for "no."

All words were presented in Chinese in the NL block (e.g., "壓歲錢") and in Japanese in the FL block (e.g., "おとしだま"). Each picture appeared twice in the same block. Once with the semantic matched word (e.g., red envelope and "New Year Envelope"), the other time with a semantic unmatched word (e.g., red envelope with "rice cake"). Same in the Japanese block. There were a total of 124 trials for each participant (Fig. 6.4).

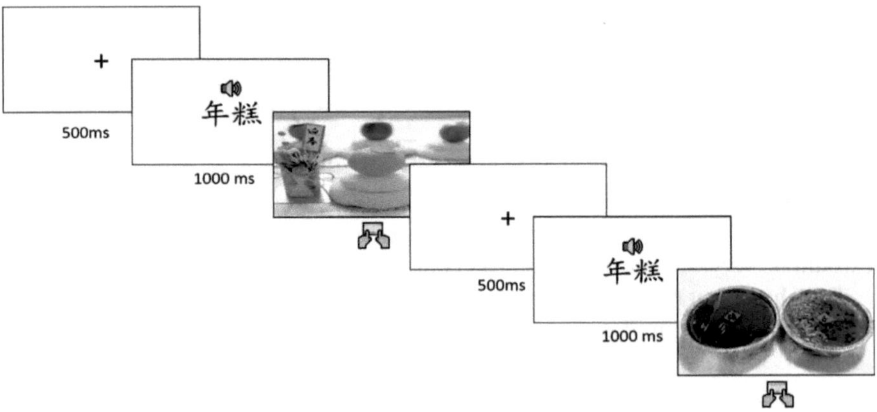

Fig. 6.4 Procedure of cultural congruency judgment in experiment 2. *Note* Two examples of cultural congruency judgment used in the Chinese block. Each trial contains 3 frames: a fixation (500 ms), a word (1000 ms), and a picture. The answer is "no" for the first and "yes" for the second

6.6 Results

6.6.1 Reaction Time

Table 6.3 and Fig. 6.5 are the results of RT in Experiment 2. A three-way ANOVA showed a significant main effect of Language $F(1, 38) = 12.56, P = 0.001, \eta_{p^2} = 0.248$; Culture Congruence also showed a significant main effect $F(1, 38) = 2.56, P = 0.14, \eta_{p^2} = 0.056$); Semantic matching had no effect. There was a significant two-way interaction of Language and Culture $F(1, 38) = 28.42, P < 0.001, \eta_{p^2} = 0.428$; Culture and Semantic Matching had a significant interaction as well $F(1, 38) = 6.38, P = 0.02, \eta_{p^2} = 0.144$. Three-way interaction was also significant $F(1, 38) = 7.12, P = 0.01, \eta_{p^2} = 0.158$.

Table 6.3 Mean RT (in msec) in experiment 2 by language, cultural congruence, and semantic matching

	Chinese		Japanese	
	Taiwanese pictures	Japanese pictures	Taiwanese picture	Japanese pictures
Semantic matched	776.3 (140.1)	841.3 (194.3)	980.7 (281.4)	921.2 (231.9)
Semantic unmatched	804.8 (150.5)	798.0 (153.8)	941.6 (237.6)	877.8 (164.3)

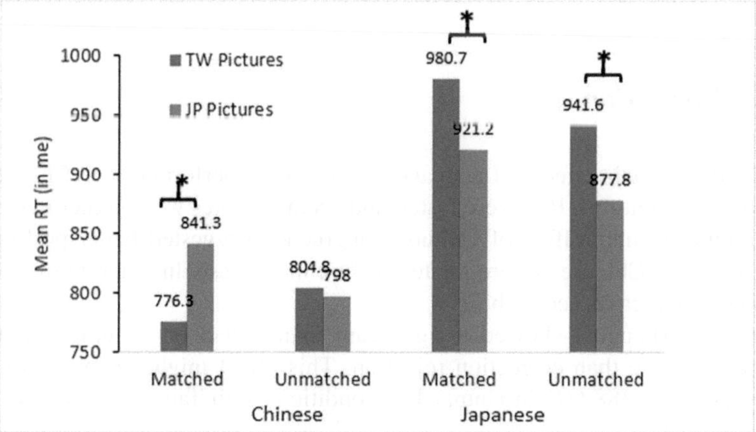

Fig. 6.5 Reaction Time (in msec) for cultural congruency judgment Task in experiment 2. *Note* The RTs for judging whether the words presented in Chinese or Japanese semantically match the pictures of Taiwanese or Japanese cultures

Table 6.4 Accuracy rate (%) for cultural congruency judgment task in experiment 2

	Chinese		Japanese	
	Taiwanese pictures	Japanese pictures	Taiwanese picture	Japanese pictures
Semantic matched	98.1 (2.5)	91.2 (7.11)	88.6 (7.6)	90.0 (7.4)
Semantic unmatched	96.7 (3.8)	97.1 (3.2)	96.9 (3.3)	96.0 (4.8)

6.6.2 *Accuracy*

Table 6.4 shows the accuracy in Experiment 2. A three-way ANOVA shows a significant main effect of Language $F(1, 38) = 24.94, p < 0.001, \eta_{p^2} = 0.396$. Accuracy in Chinese ($M = 96.55, SD = 5.22$) is higher than in Japanese ($M = 92.89, SD = 7.02$). Culture is also significant $F(1, 38) = 9.85, p = 0.003, \eta_{p^2} = 0.206$. Accuracy in Taiwanese culture ($M = 95.09, SD = 6.03$) is higher than that of Japanese culture ($M = 93.60, SD = 6.57$). Significant Semantic Matching $F(1, 38) = 45.64, p < 0.001, \eta_{p^2} = 0.546$ shows Matched condition ($M = 91.99, SD = 7.40$) has lower accuracy rate than Unmatched condition ($M = 96.70$, SD $= 3.83$). All three two-way interactions are significant. Language × Culture shows $F(1, 38) = 21.52, p < 0.001, \eta_{p^2} = 0.361$; Language × Matching shows $F(1, 38) = 32.4, p < 0.001, \eta_{p^2} = 0.460$; and Culture × Matching shows $F(1, 38) = 5.10, p = 0.03, \eta_{p^2} = 0.118$. Three-way interaction is also significant $F(1, 38) = 20.22, p < 0.001, \eta_{p^2} = 0.347$ (Fig. 6.6).

6.6.3 *Discussion*

The significant main effect of Language showed better performance in Chinese than in Japanese conditions. RTs were faster, and accuracy rates were higher in Chinese. The significant main effect of Culture Congruence suggested better performance when judging a Chinese culture-related picture in Chinese. In other words, cultural dependency increases accessibility.

Semantic Matching showed a significant main effect on accuracy, indicating a higher hits rate than correction rejection. This effect might mainly come from the low accuracy (88.6%) in Unmatched conditions with Taiwanese pictures in the Japanese block.

All two-way interactions on RT and accuracy were significant, as was the three-way interaction. In the Japanese block, the Matched conditions were slow and less accurate. This was consistent with the result in Experiment 1, in that hits required more cognitive capacity because they required an accessibility check. In the Chinese

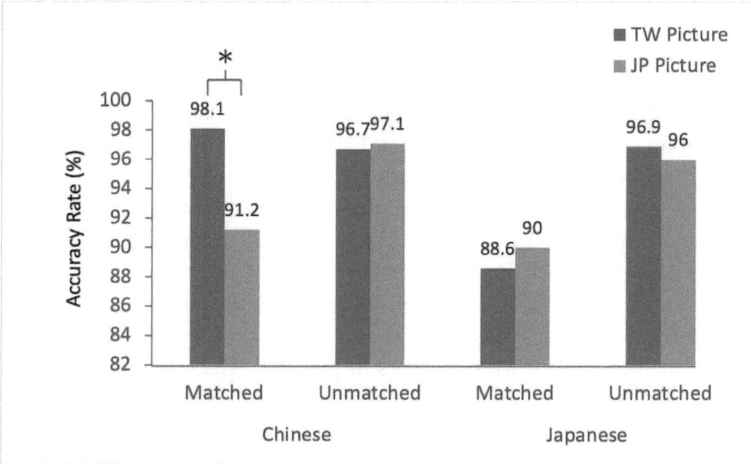

Fig. 6.6 Accuracy (%) for cultural congruency judgment task in experiment 2. *Note* The RTs were asked to judge whether the words presented in Chinese or Japanese semantically matched the pictures of Taiwanese or Japanese cultures

block, the Unmatched conditions were faster (804.8 and 798 ms) than their Japanese counterparts even though the accuracy rates were at the same level.

The most interesting result came from the Matched conditions in the Chinese block. A significant Language Congruency interaction was found. The Unmatched condition shows significant difference in accuracy rates between Taiwanese and Japanese pictures (98.1% vs. 91.2%). It suggested more difficulty in saying the correct "no" answer when receiving a Japanese picture accompanied by its Chinese translation. Take the first trial in Fig. 6.4 as an example. Participants saw the Chinese word "年糕" and a Japanese New Year rice cake picture. It took longer (841.3 ms) and was quite difficult (91.2 %) to decide whether the word accurately describes the picture. On the contrary, it only took 776.3 ms but yielded a high accuracy (98.1%) to make the decision if that was a Taiwanese New Year rice cake picture that appeared after the word.

6.7 General Discussion

Accessibility is how easily information is retrieved under certain contexts. Literature showed that both emotional involvement and the dual-process model were introduced to explain FLe. The results, however, were not consistent. Accessibility was adopted in more recent literature to assume the difference between decisions made in NL and FL is due to what information participants can retrieve to meet the task command.

A simple Semantic Matching Judgment Task was used in the two experiments in the current study. In Experiment 1, Chinese-Japanese-English trilinguals performed this semantic matching task in three language blocks. They would only see pictures of Taiwanese culture and Chinese words in the Chinese block. Same in the other two language blocks. Participants were balanced or receptive bilinguals in both Chinese-Japanese and Chinese-English tests; however, they had much higher familiarity with Japanese culture than with US culture. Not surprisingly, their performance was the best in NL condition. The matched conditions were slower and less accurate, indicating that hit responses require confirmation and need more accessibility. Finally, the better performance in the Japanese block was due to the higher familiarity and accessibility than that of the English block.

In Experiment 2, participants needed to complete the same Semantic Matching Judgment Task but only in Chinese and Japanese because a new variable, Congruency, was brought in. The stimuli were 32 pairs of pictures. The two pictures in each pair shared the same concept (e.g., New Year rice cake) but worked differently (how and when to eat them) in Taiwanese and Japanese cultures. The basic results show better performance in the NL condition than in the FL condition. Semantically Unmatched conditions were faster and more accurate. This was consistent with Experiment 1, meaning hits required more information confirmation than correct rejection. The interaction in the Matched Incongruent conditions in the Chinese block demonstrated that more accessibility was required to correctly say "no" to a Japanese picture when it was accompanied by a Chinese word, even if this Chinese word was the closest to its right meaning in Japanese culture.

The asymmetrical results between Japanese and English blocks in Experiment 1 allow us to tease apart the effect of language proficiency and familiarity on semantic decisions. The asymmetrical results between Japanese and Chinese blocks in Experiment 2 suggested that the cultural context serves as the framework for the participants to retrieve information in a task as simple as semantic matching judgment.

References

Caldwell-Harris, C. L. (2015). Emotionality differences between a native and foreign language implications for everyday life. *Current Directions in Psychological Science, 24*(3), 214–219.

Caldwell-Harris, C. L., & Ayçiçeği-Dinn, A. (2009). Emotion and lying in a non-native language. *International Journal of Psychophysiology, 71*(3), 193–204. https://doi.org/10.1016/j.ijpsycho.2008.09.006

Chan, Y. L., Gu, X., Ng, J. C. K., & Tse, C. S. (2016). Effects of dilemma type, language, and emotion arousal on utilitarian vs deontological choice to moral dilemmas in Chinese- English bilinguals. *Asian Journal of Social Psychology, 19*(1), 55–65. https://doi.org/10.1111/ajsp.12123

Cipolletti, H., McFarlane, S., & Weissglass, C. (2016). The moral foreign-language effect. *Philosophical Psychology, 29*(1), 23–40. https://doi.org/10.1080/09515089.2014.993063

Corey, J. D., Hayakawa, S., Foucart, A., Aparici, M., Botella, J., Costa, A., & Keysar, B. (2017). Our moral choices are foreign to us. *Journal of Experimental Psychology-Learning Memory and Cognition, 43*(7), 1109–1128. https://doi.org/10.1037/xlm0000356

Costa, A., Foucart, A., Arnon, I., Aparici, M., & Apesteguia, J. (2014a). Piensa" twice: On the foreign language effect in decision making. *Cognition, 130*(2), 236–254. https://doi.org/10.1016/j.cognition.2013.11.010

Costa, A., Foucart, A., Hayakawa, S., Aparici, M., Apesteguia, J., Heafner, J., & Keysar, B. (2014b). Your morals depend on language. *PLoS ONE, 9*(4), Article e94842.

Ellis, C., Thierry, G., Vaughan-Evans, A., & Jones, M. W. (2018). Languages flex cultural thinking. *Bilingualism-Language and Cognition, 21*(2), 219–227. https://doi.org/10.1017/s1366728917000190

Gao, S., Zika, O., Rogers, R. D., & Thierry, G. (2015). Second language feedback abolishes the "hot hand" effect during even-probability gambling. *Journal of Neuroscience, 35*(15), 5983–5989. https://doi.org/10.1523/JNEUROSCI.3622-14.2015

Geipel, J., Hadjichristidis, C., & Surian, L. (2015a). The foreign language effect on moral judgment: The role of emotions and norms. *PLoS ONE, 10*(7), Article e0131529. https://doi.org/10.1371/journal.pone.0131529

Geipel, J., Hadjichristidis, C., & Surian, L. (2015b). How foreign language shapes moral judgment. *Journal of Experimental Social Psychology, 59*, 8–17. https://doi.org/10.1016/j.jesp.2015.02.001

Geipel, J., Hadjichristidis, C., & Surian, L. (2016). Foreign language affects the contribution of intentions and outcomes to moral judgment. *Cognition, 154*, 34–39. https://doi.org/10.1016/j.cognition.2016.05.010

Greene, J. D. (2014). Beyond point-and-shoot morality: Why cognitive (neuro) science matters for ethics*. *Ethics, 124*(4), 695–726.

Greene, J. D., Nystrom, L. E., Engell, A. D., Darley, J. M., & Cohen, J. D. (2004). The neural bases of cognitive conflict and control in moral judgment. *Neuron, 44*(2), 389–400. https://doi.org/10.1016/j.neuron.2004.09.027

Greene, J. D., Sommerville, R. B., Nystrom, L. E., Darley, J. M., & Cohen, J. D. (2001). An fMRI investigation of emotional engagement in moral judgment. *Science, 293*(5537), 2105–2108. https://doi.org/10.1126/science.1062872

Hadjichristidis, C., Geipel, J., & Savadori, L. (2015). The effect of foreign language in judgments of risk and benefit: The role of affect. *Journal of Experimental Psychology. Applied, 21*(2), 117–129. https://doi.org/10.1037/xap0000044

Harris, C. L., Ayçiçeği, A., & Gleason, J. B. (2003). Taboo words and reprimands elicit greater autonomic reactivity in a first language than in a second language. *Applied Psycholinguistics, 24*(04), 561–579.

Hayakawa, S., Tannenbaum, D., Costa, A., Corey, J. D., & Keysar, B. (2017). Thinking more or feeling less? Explaining the foreign-language effect on moral judgment. *Psychological Science, 28*(10), 1387–1397. https://doi.org/10.1177/0956797617720944

Kahneman, D. (2003). A perspective on judgment and choice: Mapping bounded rationality. *American Psychologist, 58*(9), 697–720. https://doi.org/10.1037/0003-066x.58.9.697

Keysar, B., Hayakawa, S. L., & An, S. G. (2012). The foreign-language effect: Thinking in a foreign tongue reduces decision biases. *Psychological Science, 23*(6), 661–668. https://doi.org/10.1177/0956797611432178

Knez, I., & Nordhall, O. (2017). Guilt as a Motivator for Moral Judgment: An Autobiographical Memory Study. *Frontiers in Psychology, 8*, Article 750. https://doi.org/10.3389/fpsyg.2017.00750

Körner, A., Tscharaktschiew, N., Schindler, R., Schulz, K., & Rudolph, U. (2016). The everyday moral judge—Autobiographical Recollections of Moral Emotions. *PLOS ONE, 11*(12), Article e0167224. https://doi.org/10.1371/journal.pone.0167224

Kusev, P., van Schaik, P., & Aldrovandi, S. (2012). Preferences induced by accessibility: Evidence from priming. *Journal of Neuroscience, Psychology, and Economics, 5*(4), 250–258. https://doi.org/10.1037/a0030289

Li, P., Legault, J., & Litcofsky, K. A. (2014). Neuroplasticity as a function of second language learning: anatomical changes in the human brain. *Cortex, 58*, 301–324. https://doi.org/10.1016/j.cortex.2014.05.001

Malt, B. C., Li, P., Pavlenko, A., Zhu, H., & Ameel, E. (2015). Bidirectional lexical interaction in late immersed Mandarin-English bilinguals. *Journal of Memory and Language, 82*, 86–104. https://doi.org/10.1016/j.jml.2015.03.001

Marian, V., & Neisser, U. (2000). Language-dependent recall of autobiographical memories. *Journal of Experimental Psychology: General, 129*(3), 361–368. https://doi.org/10.1037/0096-3445.129.3.361

Matsumoto, A., & Stanny, C. (2006). Language-dependent access to autobiographical memory in Japanese-English bilinguals and US monolinguals. *Memory, 14*(3), 378–390. https://doi.org/10.1080/09658210500365763

Miozzo, M., Navarrete, E., Ongis, M., Mello, E., Girotto, V., & Peressotti, F. (2020). Foreign language effect in decision-making: How foreign is it? *Cognition, 199*, Article 104245. https://doi.org/10.1016/j.cognition.2020.104245

Mortensen, L., Berntsen, D., & Bohn, O.-S. (2015). Retrieval of bilingual autobiographical memories: Effects of cue language and cue imageability. *Memory, 23*(2), 138–156. https://doi.org/10.1080/09658211.2013.873809

Muda, R., Niszczota, P., Bialek, M., & Conway, P. (2018). Reading dilemmas in a foreign language reduces both deontological and utilitarian response tendencies. *Journal of Experimental Psychology-Learning Memory and Cognition, 44*(2), 321–326. https://doi.org/10.1037/xlm0000447

Tversky, A., & Kahneman, D. (1981). The framing of decisions and the psychology of choice. *Science, 211*(4481), 453–458.

Wu, Y. J., & Thierry, G. (2010). Chinese-English bilinguals reading English hear Chinese. *Journal of Neuroscience, 30*(22), 7646–7651. https://doi.org/10.1523/JNEUROSCI.1602-10.2010

Wu, Y. J., & Thierry, G. (2012). How reading in a second language protects your heart. *Journal of Neuroscience, 32*(19), 6485–6489. https://doi.org/10.1523/JNEUROSCI.6119-11.2012

Chapter 7
Reciprocal Relationship Between Statistical Learning and Chinese Character Recognition

Denise Hsien Wu, Andhika Renaldi, and Yi-Syuan Huang

Abstract Extraction of structural regularity embedded in the orthography of any language is the foundation for learning to read. Recent findings from our and other laboratories have accumulated to show that such ability, as reflected in indices of statistical learning (SL), contributes to the processing and acquisition of reading a native or a foreign language. On the other hand, whether experience with specific kinds of regularity through reading behaviors modulates the corresponding ability of SL has rarely been explored. In this chapter, we report evidence from a novel SL paradigm that measures the sensitivity to temporal and spatial positional regularity. Specifically, the stimuli in the spatial positional SL (PSL) task resemble the orthographic structure of Chinese phonograms. The results showed that individual differences in the spatial PSL performance correlated with the sensitivity to orthography of Chinese characters in foreign adults. Furthermore, Chinese readers, who have extensive exposure to Chinese characters, perform better than foreigners, particularly in the spatial but not the temporal PSL task. In addition, critical components of event-related potentials underlying the sensitivity to spatial and temporal positional regularity, which are further modulated by reading experience, are identified. The behavioral and neurophysiological evidence consistently supports the functional and neuronal distinctions between spatial and temporal PSL, and the reciprocal relationship between spatial PSL and orthographic processing of Chinese characters. These findings not only extend the theoretical considerations of SL research from alphabetic to logographic orthographies, they also highlight the importance and strengths of cross-linguistic comparisons in SL and reading studies.

D. H. Wu (✉) · A. Renaldi · Y.-S. Huang
Institute of Cognitive Neuroscience, National Central University, Taoyuan, Taiwan
e-mail: denisewu@cc.ncu.edu.tw

A. Renaldi
Taiwan International Graduate Program in Interdisciplinary Neuroscience, National Central University and Academia Sinica, Taipei, Taiwan

Y.-S. Huang
International Ph.D. Program in Interdisciplinary Neuroscience, National Central University and University System of Taiwan, Taoyuan, Taiwan

D. H. Wu (ed.), *Cognitive and Neural Foundations of Chinese Reading*, Chinese Language Learning Sciences, https://doi.org/10.1007/978-981-96-6669-0_7

Keywords Reading · Literacy acquisition · Logographic orthography · Spatial positional regularity · Electrophysiology

7.1 Statistical Learning of Different Types of Stimuli with Different Dependencies in Different Modalities

Statistical learning (SL) is an essential mechanism through which humans and other animals perceive, understand, and predict patterns in their environment. It involves gathering statistical properties from the sensory input, enabling individuals to make sense of complex patterns in various domains. Such ability is regarded as "a major component of the process by which internal representations of the environment are developed" (Fiser et al., 2010). By leveraging SL, an experienced cook can intuitively balance various flavors in dishes without relying on recipes, while an experienced sailor can predict the upcoming weather based on observations of the sky and sea conditions. In other words, the application of SL enables us to extract embedded structure in incoming information of various domains for the purposes of generalization and prediction.

In the seminal study conducted by Saffran et al. (1996), in which the term SL was first coined, eight-month-old infants and adults were found to be sensitive to triplets of syllables embedded in a two-minute auditory stream. An equally impressive SL ability in the visual modality was also reported in Fiser and Aslin (2002). Although SL has been regarded as a domain-general ability in relatively early literature (e.g., Altmann et al., 1995; Kirkham et al., 2002), accumulated evidence has challenged this assumption by demonstrating distinct performance characteristics observed in various SL tasks. One line of research contrasting SL in the auditory and visual modalities has shown different biases (Conway & Christiansen, 2005, 2009) and different optimal presentation rates (Emberson et al., 2011). Individual differences of SL in the auditory and visual modalities also exhibited no or low correlation when verbal (i.e., syllables) and nonverbal (i.e., geometric shapes) materials were employed in the SL tasks, respectively (Siegelman & Frost, 2015). Indeed, SL performance of regularity presented in verbal and nonverbal materials has been documented to differ from each other in recent literature. Thiessen (2010) demonstrated that when learning regularity embedded in a stream of paired auditory–visual stimuli, adults benefited from verbal information (i.e., syllables) in the auditory modality but not from nonverbal information (i.e., pure tones), suggesting that learners are more sensitive to the relationship between word-object than tone-object pairs. In an auditory SL task, Siegelman and colleagues (2018) reported high and low internal consistency among nonverbal and verbal test items, respectively. In a visual SL task, Perfors and Kidd (2022) also reported better performance in learning regularity presented in verbal (familiar) than nonverbal (unfamiliar) stimuli. All these findings above underscore the influences of presentation modalities and stimulus properties on SL performance.

As in our review of studies supporting domain-specificity of the SL ability (Wu & Bulut, 2020), there is increasing evidence for distinct sensitivity to different types of regularity presented in various SL paradigms. In contrast with the prevalent SL research on adjacent dependency among materials, relatively little work focuses on non-adjacent dependency. Among the early endeavors in this direction, Gomez (2002) reported that both adults and 18-month-old infants could distinguish three-elements strings from two artificial languages that only differed in non-adjacent dependency between the first and the third element, especially when the adjacent dependency was least predictable. That is, the sensitivity to the non-adjacent dependency between the first and the third element was the highest when the second element was most variable, which prompted participants to rely on the reliable non-adjacent dependency. However, in an SL study with a continuous stream of syllable triplets, Newport and Aslin (2004) found that perfect non-adjacent regularity (i.e., TP = 1) between the first and the third syllables in a triplet could not be learned. Only when the syllables were changed to phonetic segments, either consonants or vowels, were the participants able to detect and learn the regularity. The studies on individual differences also revealed no correlation between SL of adjacent and non-adjacent dependencies (Misyak & Christiansen, 2012; Siegelman & Frost, 2015).

In addition to temporal relationship, as examined in the studies mentioned above, materials in a visual SL task can associate with each other in spatial relationship. A pioneering exploration of such spatial contingency was carried out by Fiser and Aslin (2001). Specifically, 12 shapes were organized into six base pairs, two of which were in one of three orientations (horizontal, vertical, and oblique). During the study phase, participants were exposed to a continuous stream of scenes with three base pairs presented in a 3 × 3 grid. In the test phase, participants made selections between the correct base pair and the non-base pair in two-alternative-forced-choice (2AFC) questions. The results showed that participants learned the absolute and relative positions of two shapes within a base pair rapidly and automatically. In following studies that compared the SL performance of temporal and spatial contingencies directly, it has been consistently found that regularity in the spatial format is easier to learn than regularity in the temporal format (Conway & Christiansen, 2009; Saffran, 2002). However, Deocampo et al. (2019) reported comparable SL performance between learning spatial and temporal verbal sequences, regardless of adjacent and non-adjacent dependencies. All these findings highlight the multifaceted nature of SL, with performance varying across different stimulus types, presentation modalities, and dependency formats.

7.2 Statistical Learning is Associated with Language Performance

Given that there are abundant "hidden rules" in the language system, it is natural to assume that the sensitivity to linguistic regularities, which is reflected in the efficiency of SL, contributes to the acquisition of language. Indeed, empirical evidence has demonstrated the association between SL and language performance (see Wu & Bulut, 2020, for a review). For example, individuals who excelled in discerning statistical relationships within an auditory sequence exhibited a superior ability to predict the final words of sentences in noisy speech (Conway & Christiansen, 2009). In contrast, children with specific language impairment (SLI) exhibited inferior performance in learning the transitional probabilities (TPs) of sounds presented in the background than control participants matched for age and nonverbal IQ. Moreover, the SL performance of both the SLI and control groups was correlated with individuals' vocabulary knowledge (Evans et al., 2009). Misyak and colleagues (2010, 2012) also showed that individual differences in SL performance measured by different versions of artificial grammar learning (AGL) tests correlated significantly with individual differences in sentence comprehension. Consistently, the accuracy of visual SL tests has been found to correlate with the performance in the vocabulary and grammaticality judgment tests (Daltrozzo et al., 2017).

In addition to the correlational research, findings from developmental studies have provided support for the causal contribution from SL to language acquisition: 6- and 8-month-old infants were found to be sensitive to the statistical distribution of speech sounds, and this sensitivity affected their phonetic discrimination (Maye et al., 2002). Ellis et al. (2013) reported that the performance of 6-month-old infants in a visual expectancy paradigm predicted their receptive and productive vocabulary when they turned 22 months old. In another longitudinal study, visual rule learning performance of 7-month-old infants predicted their lexical and grammatical skills at 2 years old (Bettoni et al., 2020).

Among the different facets of language processing, the link between SL and reading is of particular interest in this chapter. In normal populations, the reading ability of children and adults has been found to correlate with their performance in segmenting visual triplets in a continuous stream (Arciuli & Simpson, 2012). The neuropsychological literature on Developmental Dyslexia (DD) has also implicated a deficit in SL. For example, dyslexics have been shown to have specific "anchoring" deficits in using cross-trial statistics of pure tones for pitch discrimination, and the degree of difficulty was highly correlated with the degree of their reading difficulties (Ahissar, 2007). Pavlidou et al. (2010) also reported that children with DD failed to show implicit SL in an AGL task. The connection between deficient SL and reading difficulties has been observed even in adults with DD. Gabay et al. (2015) found that English-speaking university students diagnosed with DD had lower accuracy in linguistic and non-linguistic auditory SL tests compared to control participants. Sigurdadottir et al. (2017) also found poorer performance of Icelandic adults with DD than the control group in recognizing pairs of shapes in a visual SL test. Taken together, people diagnosed with DD at different ages and with different native languages seem to have general deficits in SL across modalities and across SL paradigms.

7.3 Statistical Learning Contributes to Learning to Read Chinese Characters

Among the endeavors to determine the contribution from SL to reading abilities, few attempts have focused on learning to read a foreign orthography. One pioneering study conducted by Frost et al. (2013) examined the improvements of word naming and nonword decoding, as well as the morphological and semantic priming effects, in American teenagers who were beginning learners of Hebrew during their first year in an immersive language learning environment. The results clearly indicated that individual differences in a visual SL task of shape-triplet segmentation correlated significantly with the enhanced sensitivity to Hebrew word morphology, which resulted in improved word naming and nonword decoding, but not to semantics. A similar attempt to determine the relationship between SL and reading acquisition of a foreign orthography is made in our laboratory: A battery of cognitive tests, including auditory and visual triplet segmentation, was administered to foreign students who came to Taiwan for college or advanced education. After controlling for the effects of acquisition time of Chinese, IQ and working memory (WM), the performance of the visual SL task adopted from Frost et al. (2013) significantly correlated with the number of Chinese characters that they could recognize (Wu & Bulut, 2020).

The significance of the association between SL and learning to read Chinese characters is that Chinese orthography is logographic in nature, which is drastically different from alphabetic orthography. Specifically, Chinese characters are the smallest orthographic units that map onto phonological units in the Chinese writing system, while the strokes or radicals do not correspond to sub-syllabic phonemes. Furthermore, the mapping between the orthography (characters) and phonology (syllables) is not random or by rote memory but presents semi-regularity and semi-consistency between sub-character phonetic radicals and character pronunciations. The majority (~70%) of Chinese characters are phonograms (e.g., 媽, which means "mother" with the pronunciation "ma1"), with a horizontal structure composed of a semantic radical predominantly on the left side (e.g., 女, which means "female" with the pronunciation "nu3") and a phonetic radical predominantly on the right side (e.g., 馬, which means "horse" with the pronunciation "ma3"). An analysis based on the Academia Sinica Balanced Corpus (Huang & Chen, 1998) indicates that there are about 700 phonetic radicals and 200 semantic radicals within the modern Chinese writing system. The phonetic and semantic radicals serve as functional cues, the former offering insights into a character's pronunciation while the latter into its meaning, respectively. This distinction between phonetic and semantic radicals plays a crucial role in the process of character recognition. Building upon this foundational understanding, we have conducted a study with magnetoencephalography (MEG) to further investigate the cognitive processing underlying recognition of Chinese characters (Hung et al., 2014). Our findings revealed that the repetition of a phonetic radical across two sequentially presented characters attenuated the early MEG responses (i.e., M170 and M250) elicited by the introduction of the second character. This phenomenon was not mirrored in the processing of semantic radicals, suggesting a prioritization or precedence of phonetic information in the early stages of Chinese character recognition.

Given the structure of Chinese orthography introduced above, it is reasoned that individuals with superior visual SL ability are sensitive to the regularity/consistency of Chinese phonograms and are able to acquire Chinese characters efficiently. To test this hypothesis directly, we have combined behavioral measurements of the visual SL ability with neuroimaging measurements to identify the brain regions that are sensitive to the structure of Chinese orthography (Yu et al., 2019). Specifically, native readers and foreign learners of Chinese performed a pronunciation verification task to determine whether a Chinese syllable presented in the auditory modality is a possible pronunciation of a pseudo-phonogram (with the legal combination of a semantic and a phonetic radical, but does not form a meaningful character) presented in the visual modality. Both groups of participants reached consistent judgments, which were assumed to rely on the application of the same set of orthography-to-phonology conversion (OPC) rules emerging from sufficient encounters of real Chinese characters. Through the analysis of the fMRI results, we first confirmed the involvement of a left-lateralized neural network, including left inferior-temporal-parietal brain regions, for the OPC processing of character recognition in both native and non-native readers of Chinese. Next, we took the neural network as regions of interests (ROIs) to search for the neural correlates of the sensitivity to TPs among geometric shapes in a visual SL test as employed in Frost et al. (2013). For non-native readers, the correlation analysis revealed individuals' activations in the left inferior and middle frontal gyri to be negatively correlated with their accuracy in the visual SL test. For native readers, on the other hand, the same analysis identified the activations in posterior brain regions, including the left parietal lobule and bilateral fusiform gyri, to be associated with the visual SL scores. These findings suggest that as the proficiency of Chinese character recognition increases, the relevant load of OPC processing, which is reflected in the activations of various critical brain regions in the reading network, decreases in both native and nonnative readers.

In parallel with our findings (Wu & Bulut, 2020; Yu et al., 2019), Tong and colleagues (2019) have also provided evidence for the link between visual SL abilities and awareness of Chinese orthography in children in Hong Kong. Specifically, Grade-2 children with DD performed worse than chronologically age-matched controls in VSL, as well as in word reading and in orthographic awareness. In addition, the performance in VSL was significantly associated with that in word reading and orthographic awareness in children with DD. Although these recent findings are correlational, hence cannot argue for a causal contribution from SL to the extraction of OPC unambiguously, they do provide strong support for the critical role of SL in reading of a non-alphabetic orthography.

7.4 Sensitivity to Chinese Orthographic Structure Correlates Particularly with Statistical Learning of Spatial Positional Regularity

To investigate the association between SL and language performance, stimuli with temporal contingency are employed in most of such studies. Indeed, there is abundant temporal regularity embedded in multiple levels of natural language. Take English for example, the letter cluster "un," which bears the meaning of "not" as a prefix, is more likely to appear at the beginning than the end of a word, while the letter string "ness," which bears the meaning of "noun quality/condition" as a suffix, has the opposite likelihood distribution in a word. Because of the close correspondence between orthography and phonology in alphabetic language, which is sequentially represented in time, the statistical distribution of letter clusters illustrated above is a kind of temporal regularity in nature. Similarly, the morphosyntactic dependency (e.g., subject-noun and verb agreement) and the canonical word order (e.g., subject-verb-object) also demonstrate a kind of temporal regularity at the sentence level. Therefore, it may not be surprising to find that sensitivity to TPs in temporal displays correlates with the proficiency of language processing.

Chinese is similar to English and other languages in exhibiting temporal regularity at the morphological and syntactic levels. For example, the character "這," which bears the meaning of "this," is more likely to appear at the beginning than the end of a multi-character word, while the character "者," which bears the meaning of "person" (similar to "-er" in English), has the opposite distribution likelihood in a multi-character word. At the sentence level, the aspectual agreement across words in a distance also reflects a kind of temporal regularity (e.g., 麵包師傅已經預備了甜點), and its violation (e.g., 麵包師傅正在預備了甜點) has been found to elicit strong magnitude of a neuromagnetic component as an index of syntactic processing (Tsai et al., 2013). Moreover, correct comprehension of the meaning of sentences that are composed of the same set of characters with different orders (e.g., 民眾討厭那個毆打明星的攝影師 vs. 民眾討厭那個明星毆打的攝影師) has been shown to require different reading time (e.g., Xu et al., 2019) and different neuronal resources (e.g., Bulut et al., 2018) depending on the processing difficulty. All these findings suggest that sensitivity to temporal information embedded in linguistic materials may play an important role to reading Chinese.

In addition to temporal regularity, however, there is also spatial regularity in the Chinese writing system. The fundamental difference between alphabetic (such as English) and non-alphabetic (such as Chinese) orthography is the correspondence between orthography and phonology. An alphabetic word consists of a sequence of letters that represent graphemes corresponding to phonemes, while a logographic Chinese character consists of spatial stroke patterns that form radicals. As introduced above, the phonetic radical, which provides a cue to the syllable pronunciation of a phonogram, usually appears on the right side of the character. Interestingly, the same radical can serve different functions depending on its position within a phonogram. Take the radical "女" for example. Its meaning is "female" when standing alone as a character, and is pronounced as "nu3." This radical is more likely to appear on the left side of a character (e.g., 媽, which means "mother" and is pronounced as "ma1") than on the right side of a character (e.g., 汝, which means "you" and is pronounced as "ru3"). That is, when the radical "女" appears on the left as a semantic radical, it denotes the semantic category of the whole character. In contrast, when it appears on the right as a phonetic radical, it rhymes with the syllable pronunciation of the whole character. In addition, the distribution statistics of different radicals would be accessed and processed simultaneously when a character is encountered. In other words, the sensitivity to semantic and phonetic radicals within Chinese characters is spatial but not temporal in nature.

Although there have been few attempts to examine SL of spatial contingencies as relative positions between stimuli (e.g., Fiser & Aslin, 2001), the sensitivity to stimuli appearing at specific spatial positions was only explored in one study by Chetail (2017). Specifically, a sequence of five nonverbal symbols resembling Latin alphabets was presented visually to native French participants who do not have prior knowledge of the stimuli. During the study phase, pairs of two symbols adjacent to each other (i.e., a bigram) always occupied fixed spatial positions. After several minutes of exposure, participants were able to distinguish stimuli with frequent bigrams from novel stimuli. More importantly, participants were also able to distinguish stimuli with a bigram in the familiarized positions from those with a bigram in unfamiliarized positions. In other words, participants not only learned the association between two symbols but also their spatial positions within the five-symbol sequence.

To explore the sensitivity to probabilities of non-linguistic stimuli appearing at specific spatial positions, we have developed and optimized a novel visual SL task whose stimuli are geometric shapes. Critically, each shape only appeared at the left or right spatial position of a horizontally aligned pair, resembling the semantic and phonetic radicals in the most common types of Chinese phonograms. In parallel with the SL task that presents spatial positional regularity (spatial PSL), a visual SL task whose stimuli only appear in the first or second temporal position of a consecutively aligned pair (temporal PSL) has also been developed (Renaldi, unpublished master thesis). Specifically, 40 geometric shapes that are difficult to name verbally were prepared and divided into two sets to be employed in the temporal and spatial VSL tasks separately. Among the 20 shapes employed in each of the VSL tasks, 16 shapes were used in a study phase (i.e., "studied" shapes), while the remaining four shapes were preserved to be only used in a testing phase (i.e., "novel" shapes). Eight of the studied shapes always appeared in the first and the left position in the temporal and the spatial PSL test, respectively. On the other hand, the other eight studied shapes always appeared at the second and the right position in the temporal and the spatial PSL test, respectively. In the study phase, participants were exposed to the temporal and spatial pairs of stimuli for about nine minutes.

In the testing phase of both the temporal and spatial PSL tasks, 2- and 4-alternative-forced-choice questions of generalization judgment and pattern completion were prepared. For each question, the correct answer (i.e., a studied shape in its regular temporal/spatial position with a novel shape in the other position within a pair) was presented with the foil alternative (i.e., a studied shape in the opposite position to its regular position with the same novel shape in the other position within a pair). Because the novel shape never appeared in the study phase, participants could not perform the task correctly based on rote memory (or familiarity) with the pairs that were encountered before. Rather, participants had to rely on "positional rules" of the shapes to make a correct response.

The PSL tasks were administered to Taiwanese students who were proficient readers of Chinese and foreign students whose Chinese vocabulary size ranged from 0 to 438 characters. A modified lexical decision task in which participants rated 50 Chinese real characters, 25 pseudo-characters (i.e., characters composed of legal radicals in their conventional spatial positions within characters), and 25 non-characters (i.e., characters composed of legal radicals in unconventional spatial positions or of stroke patterns that do not form legal radicals) was also administered. The mean rating difference between the real character and non-character conditions generally reflects participants' knowledge of Chinese characters, while the mean rating difference between the pseudo-character and non-character conditions reflects purely participants' sensitivity to Chinese orthography. In addition, a battery of cognitive tests, including non-verbal IQ and working memory, was also administered.

The results from our previous studies demonstrated that the spatial and temporal positional regularity presented in the novel paradigm is learnable generally, as the group performance exceeded the chance level, despite great individual differences. Such findings were replicated, and the test–retest reliability of the PSL tasks was also verified (Renaldi & Wu, manuscript in preparation). Importantly, although the performance in the spatial and temporal PSL tasks correlated with each other significantly, they seemed to reflect at least partially separate mechanisms, as only accuracy of spatial but not temporal PSL correlated with foreign participants' sensitivity to Chinese orthography. Such findings not only indicate the functional distinction between spatial and temporal PSL, but also support the relationship between SL, particularly the sensitivity to spatial positional regularity, and Chinese character processing.

7.5 Reading Experience Modulates the Ability of Statistical Learning

In the literature on reading research, it has been extensively documented that proficient readers are sensitive to the regularity/consistency embedded in the writing system (Coltheart et al., 2001; Seidenberg & McClelland, 1989). For example, lexical decision and naming of the regular word "mint" are faster and more accurate than the same responses on the irregular word "pint." Words/nonwords with many orthographic neighbors, such as "sand," are also read aloud faster than words/nonwords with few or no orthographic neighbors, such as "club" (see Andrews, 1997, for a review). The consistency and regularity effects highlight how the predictability of OPC in a language influences the ease with which readers can decode words

(Ziegler & Goswami, 2005). Moreover, such knowledge is reported to develop early in young ages. Cassar and Treiman (1997) found that first-grade native English speakers could distinguish non-words ending with allowable doublets (e.g., "baff") as more word-like than those starting with doublets (e.g., "bbaf") and those ending with unallowable doublets (e.g., "yihh"). Similar findings were replicated in young native speakers of French (Pacton et al., 2001) and of German (Rothe et al., 2015).

Although Chinese orthography differs fundamentally from alphabetic orthographies as introduced, there is ample evidence showing that Chinese readers are sensitive to the orthographic regularity presented in the radicals within a character. For example, responses on characters with higher consistency, such as "情 qing2," who has several orthographic neighbors (清 qing1, 晴 qing2, 睛 jing1) that share the same phonetic radical "青" and rhymed pronunciations, have been found to be more accurate and faster than responses on characters with lower consistency, such as "梳 shu1" whose orthographic neighbors are "疏 shu1," "流 liu2," and "毓 yu4" (e.g., Lee et al., 2005; Yang et al., 2009). The developmental trajectory of such a consistency effect has also been documented in beginning readers in the elementary school (Tzeng & Lee, 2012). Ho and Bryant (1997) found that first- and second-graders relied on phonetic radicals to name Chinese characters, exhibiting a regularity effect. Ho and colleagues (2003) also reported the knowledge, though not fully mature, of the position, function, and sound value of phonetic radicals in first graders. On the other hand, the knowledge of the position and semantic category of semantic radicals also seemed to develop early in first graders, but the knowledge of the function of semantic radicals was not reliably observed until the third grade. Furthermore, several measures of the knowledge of semantic and phonetic radicals correlated with participants' reading performance. These findings underline the link between sensitivity to orthographic structure and reading ability.

Despite the demonstrations of readers of different orthographies being sensitive to the regularity specific to their language experience, few attempts have been made to investigate the effects of linguistic inputs on SL performance directly. One exception is the study conducted with Hebrew-speaking participants (Siegelman et al., 2018), which aimed at explaining the low or no correlation between SL performance in auditory and visual modalities in previous literature (e.g., Siegelman & Frost, 2015). Specifically, it was hypothesized that learners' performance in an auditory SL task, in which verbal materials are typically employed, is not only determined by what they learn from novel auditory input, but is also affected by their "entrenched expectations about speech co-occurrences from their native language" (p. 198, Siegelman et al., 2018). On the other hand, learners' performance in a visual SL task, in which nonverbal materials are typically employed, might not be under any systematic influence of prior knowledge or experience. The results from this study provide strong support for the hypothesis, as the SL performance in a visual nonverbal task only correlated with that in an auditory nonverbal task, but not with that in an auditory verbal task. In addition, the learning accuracy of stimuli that are more "Hebrew-like" in the auditory verbal task was higher than that of less "Hebrew-like" stimuli. These findings clearly indicate that extensive language experience enhances the sensitivity to regularity akin to the structure embedded in linguistic inputs.

Taking the "entrenchment hypothesis" of Siegelman et al. (2018) one step further, we proposed that long-time exposure to Chinese orthography would improve readers' sensitivity to spatial positional regularity as measured in the spatial PSL task, which is introduced above. To investigate this proposal directly, we have collaborated with colleagues at the Hebrew University to compare the spatial PSL accuracy of college students in Taiwan and in Israel. The results support our hypothesis in showing a trend that participants in Taiwan who are proficient readers of Chinese performed better than participants in Israel who do not read Chinese in the spatial but not in the temporal PSL task (Renaldi et al., manuscript in preparation). This trend was enhanced and significant if the spatial PSL was administered following a sound-shape association task, in which language-specific pseudo-syllables were paired with geometric shapes later to be learned in the spatial PSL task to implicitly induce participants' orthographic knowledge of their native language, after taking participants' non-verbal IQ and working memory performance into account (Renaldi et al., manuscript in preparation). Through this series of experiments, we and our colleagues have demonstrated for the first time the direct influence of different reading experiences on sharpening the sensitivity to a specific type of regularity embedded in non-linguistic stimuli.

7.6 Sensitivity to Temporal and Spatial Positional Regularity Is Reflected in Distinct Neurophysiological Components

Neuroimaging techniques with superior temporal resolution, namely, event-related potentials (ERPs), have been applied to investigate brain activities when participants perform conventional SL tests in the past decades. One relatively early attempt of such endeavors measured and compared the ERP waveforms elicited by the onsets of pronounceable pseudowords before and after training during which adults listened to these pseudowords presented in a continuous speech stream repeatedly (Sanders et al., 2002). The results showed that for adults who demonstrated large behavioral learning effects, the magnitude of both N100 and N400 elicited by word onsets increased after training, while only the magnitude of N400 increased after training in adults who demonstrated small learning effects. In line with this early study, Abla et al. (2008) employed a similar triplet segmentation task but replaced the stimuli with continuous, non-linguistic auditory tone sequences. They reported that both N100 and N400 reflected sensitivity to the first tone (i.e., a "word onset" effect) in comparison to the third tone of triplets in high and middle performers but not in low performers. Moreover, for high performers the magnitude of N100 and N400 was negatively correlated with the transitional probabilities among tones in a triplet early in the experiment, while the same association was revealed relatively late for middle performers, but no such association was observed for low performers. Using musical stimuli, Pearce et al. (2010) also found that notes with low probability in ecologically valid melodies elicited a larger negative ERP component than notes with high probability in the time window comparable to that of N400 (i.e., between 400 and 450 ms). In the same study, EEG results also showed that low-probability notes elicited a larger beta band (14–30 Hz) oscillation and a larger long-range phase synchronization between multiple brain regions than high-probability notes over the parietal lobe (Pearce et al., 2010).

The association of N400 with SL is not only observed in adults but also in younger populations. For example, Kooijman et al. (2005) presented two-syllable words in a continuous speech stream to 10-month-old infants. It was found that the young infants already showed sensitivity to those two-syllable words that they had encountered before, as reflected in the ERP signals during the time window between 300 and 500 ms after the word onset. In the same vein, Teinonen et al. (2009) employed the word segmentation paradigm to reveal the neurophysiological correlates of the SL effect in sleeping neonates. Specifically, ERPs elicited by the first syllable in three-syllable pseudowords exhibited a more negative magnitude than the ERPs elicited by the third syllable between 260 and 440 ms (peaking at 340–390 ms) after the onset of the syllable in newborn infants after they were exposed to the pseudowords briefly (<1 h). These results strongly argue for the automaticity of SL at least in the auditory modality.

In addition to N400, other ERP components have also been associated with SL in the auditory modality. Cunilera et al. (2006) found that while the N400 component reflected online segmentation of word onsets in continuous speech, the P200 component was further modulated by stress cues presented in the speech stream. van Zuijen et al. (2006) employed an "abstract" oddball paradigm together with simultaneous ERP recording to reveal the involvement of mismatch negativity (MMN), an ERP component elicited by rare sound deviants that do not match extracted regularity, in an auditory SL test, regardless whether participants were aware of the regularity in the sound sequence. On the other hand, the P300 component of ERPs reflected participants' explicit knowledge of the regularity, as the deviants only modulated the magnitude of P300 when participants developed awareness of the regularity during training. Friederici et al. (2011) also reported that German infants as young as 4 months old showed sensitivity, as indexed by late (640–1040 ms) positive ERP waveforms, to nonadjacent dependency between verbs and suffices in spoken Italian sentences after only brief exposure to correct sentences of the same type.

Although there are fewer studies examining the neurophysiological correlates of SL in the visual than in the auditory modality, recent findings have shed light on the commonality and specificity of the time courses and the underlying mechanisms of SL across these two modalities. Consistent with their findings in the auditory modality, Abla and Okanoya (2009) reported that SL of shape triplets embedded in a continuous shape sequence was reflected in the N400 component of ERPs elicited by the first shape in triplets only in high performers who demonstrated large learning effects in a behavioral test. Such results indicated that N400 is an indicator of online segmentation of sequences in both auditory and visual modalities. Mueller et al. (2005) and Tabullo et al. (2013) also found that an enhanced magnitude of N400, as well as that of other ERP components including P600, was elicited by violations of semantic and/or grammatical regularities in one's native language or in a newly learned artificial language. However, Christiansen et al. (2012) only identified the P600 but not the N400 effect to be commonly associated with encountering ungrammatical structures in both the tasks of sequential learning and natural language processing with similar topographic distribution. On the other hand, a lateralized anterior negativity (LAN) was only associated with natural language processing but not sequential learning. Recent neurophysiological studies with a novel predictor-target paradigm, which is modified from the standard oddball paradigm, further revealed the involvement of P300 particularly over the centro-parietal electrodes in implicit SL tests (Daltrozzo et al., 2017; Jost et al., 2015). As pointed out by Daltrozzo and Conway (2014), further research is definitely needed to clarify the appearance and the nature of these ERP components in various SL tests across modalities.

As briefly mentionede, we have developed a novel SL paradigm that measures the sensitivity to positional regularity in the temporal order and the spatial location of simple geometric shapes appearing within a pair embedded in a continuous stream in the visual modality (Renaldi & Wu, manuscript in preparation). The spatial PSL task with a horizontal array of two nonverbal shapes that are simultaneously presented is designed specifically to mimic the orthography of Chinese characters, the majority of which consist of two radicals that are in a left–right composition (Hsiao & Shillcock, 2006). In parallel, the temporal PSL task with a sequential array of the same pairs of nonverbal shapes is also devised. The behavioral evidence from the PSL paradigm introduced above has already indicated the (at least partial) distinction between spatial and temporal PSL, as only the former but not the latter performance correlated with alphabetic readers' sensitivity to Chinese orthography after they were exposed to this writing system briefly (Renaldi & Wu, manuscript in preparation).

To further provide empirical evidence for the neural substrates underlying these two mechanisms, we measured participants' ERPs when they carried out the modified PSL task which included "odd-balls" that violated the spatial/temporal regularity during the familiarization phase. By concentrating on only learners in the PSL task, whose behavioral accuracy at the testing phase was above the chance level, we observed distinct neurophysiological correlates associated with SL of temporal and spatial positional regularity when comparing the ERP waveforms elicited by the stimuli that followed and violated the positional rules as regular and deviant stimuli, respectively. More interestingly, the neurophysiological correlates of only spatial but not temporal PSL ability were further modulated by participants' language background. Specifically, the sensitivity to temporal positional regularity was reflected in the MMN (marginally significant) and in the N400 (significant) components in both Taiwanese and foreign participants, while the sensitivity to spatial positional regularity was reflected in the MMN component in Taiwanese participants but in the P600 participants in foreign participants. Such findings underscore the distinction between temporal and spatial PSL, and their bi-directional relationship with reading experience.

7.7 Conclusion

The intricate interplay between SL and word recognition delineated in this chapter underscores SL's pivotal role across various modalities and dependencies, contributing fundamentally to linguistic competence and literacy. This synthesis reveals that SL, a versatile cognitive faculty, is crucial not only for acquiring native languages but also for decoding and comprehending non-native alphabetic or logographic orthographies.

For the debate of the commonality or independence of SL of different materials in different modalities, relatively recent evidence seems to support a shared mechanism underlying the computations of different types of dependencies after taking the existing knowledge/expectations of the stimuli into account (Siegelman et al., 2018). However, our behavioral and neuroimaging findings of the distinct performance associated with sensitivity to spatial and temporal positional regularity provide further indications that the general computational principles of SL are exhibited in spatial and temporal domains separately. Moreover, the processing of particular types of SL can be enhanced selectively by language experience. Although individuals proficient in one aspect of SL are likely to exhibit competencies across other types, significant correlations between different SL tasks do not necessarily imply a unified underlying mechanism. Indeed, the notable dissociation observed between performance in spatial and temporal PSL tasks underscores the presence of unique subcomponents within the general framework of SL.

In the context of reading Chinese characters, an area specifically emphasized in this chapter, SL of spatial positional regularity emerges as a particularly potent mechanism. The logographic nature of Chinese orthography, which integrates semantic and phonetic radicals, presents unique challenges and learning opportunities. Our findings suggest that individuals proficient in visual SL are better equipped to navigate the complexities of Chinese characters, benefiting from an acute sensitivity to the spatial positional regularities of these orthographic units. Insights from neurophysiological results further validate the contributions of SL to reading proficiency. Prominent ERP components such as MMN, N400, and P600 not only reflect the detection of positional regularities but also adaptively respond to the temporal and spatial dynamics of visual inputs. These neuronal markers highlight SL's role in the real-time processing and integration of nonverbal information, which resembles the spatial and temporal regularities embedded in linguistic inputs in an abstract manner. Such results facilitate a more nuanced understanding of SL contribution to literacy acquisition.

In summary, this chapter reaffirms the significance of SL as a foundational skill in learning to read. The empirical findings highlight the complexity of SL mechanisms and suggest that tailored interventions that consider these distinct computations may optimize the practices of language and literacy education. The potential for applying these insights to increase our understanding of SL, and to improve literacy in both native and foreign languages is immense, paving the way for targeted educational strategies that cater to the diverse cognitive profiles found in individual learners. It also underscores the necessity for further research to explore SL's multifaceted impact on language proficiency across different linguistic and cultural contexts, ensuring that all learners have the tools necessary to succeed in a multilingual and increasingly complex global landscape.

References

Abla, D., Katahira, K., & Okanoya, K. (2008). On-line assessment of statistical learning by event-related potentials. *Journal of Cognitive Neuroscience, 20*(6), 952–964.

Abla, D., & Okanoya, K. (2009). Visual statistical learning of shape sequences: An ERP study. *Neuroscience Research, 64*(2), 185–190.

Altmann, G., Dienes, Z., & Goode, A. (1995). Modality independence of implicitly learned grammatical knowledge. *Journal of Experimental Psychology: Learning, Memory, and Cognition, 21*(4), 899.

Ahissar, M. (2007). Dyslexia and the anchoring-deficit hypothesis. *Trends in Cognitive Sciences, 11*(11), 458–465.

Andrews, S. (1997). The effect of orthographic similarity on lexical retrieval: Resolving neighborhood conflicts. *Psychonomic Bulletin & Review, 4*(4), 439–461.

Arciuli, J., & Simpson, I. C. (2012). Statistical learning is lasting and consistent over time. *Neuroscience Letters, 517*(2), 133–135.

Bettoni, R., Riva, V., Cantiani, C., Molteni, M., Macchi Cassia, V., & Bulf, H. (2020). Infants' learning of rule-based visual sequences predicts language outcome at 2 years. *Frontiers in Psychology, 11*, Article 497718.

Bulut, T., Cheng, S. K., Xu, K. Y., Hung, D. L., & Wu, D. H. (2018). Is there a processing preference for object relative clauses in Chinese? *Evidence from ERPs. Frontiers in Psychology, 9*, 995.

Cassar, M., & Treiman, R. (1997). The beginnings of orthographic knowledge: Children's knowledge of double letters in words. *Journal of Educational Psychology, 89*(4), 631.

Chetail, F. (2017). What do we do with what we learn? Statistical learning of orthographic regularities impacts written word processing. *Cognition, 163*, 103–120.

Christiansen, M. H., Conway, C. M., & Onnis, L. (2012). Similar neural correlates for language and sequential learning: Evidence from event-related brain potentials. *Language and Cognitive Processes, 27*(2), 231–256.

Coltheart, M., Rastle, K., Perry, C., Langdon, R., & Ziegler, J. (2001). DRC: A dual route cascaded model of visual word recognition and reading aloud. *Psychological Review, 108*(1), 204.

Conway, C. M., & Christiansen, M. H. (2005). Modality-constrained statistical learning of tactile, visual, and auditory sequences. *Journal of Experimental Psychology: Learning, Memory, and Cognition, 31*(1), 24.

Conway, C. M., & Christiansen, M. H. (2009). Seeing and hearing in space and time: Effects of modality and presentation rate on implicit statistical learning. *European Journal of Cognitive Psychology, 21*(4), 561–580.

Cunillera, T., Toro, J. M., Sebastián-Gallés, N., & Rodríguez-Fornells, A. (2006). The effects of stress and statistical cues on continuous speech segmentation: An event-related brain potential study. *Brain Research, 1123*(1), 168–178.

Daltrozzo, J., & Conway, C. M. (2014). Neurocognitive mechanisms of statistical-sequential learning: What do event-related potentials tell us? *Frontiers in Human Neuroscience, 8*, 437.

Daltrozzo, J., Emerson, S. N., Deocampo, J., Singh, S., Freggens, M., Branum-Martin, L., & Conway, C. M. (2017). Visual statistical learning is related to natural language ability in adults: An ERP study. *Brain and Language, 166*, 40–51.

Deocampo, J. A., King, T. Z., & Conway, C. M. (2019). Concurrent learning of adjacent and nonadjacent dependencies in visuo-spatial and visuo-verbal sequences. *Frontiers in Psychology, 10*, 1107.
Ellis, E. M., Gonzalez, M. R., & Deák, G. O. (2014). Visual prediction in infancy: What is the association with later vocabulary? *Language Learning and Development, 10*(1), 36–50.
Emberson, L. L., Conway, C. M., & Christiansen, M. H. (2011). Timing is everything: Changes in presentation rate have opposite effects on auditory and visual implicit statistical learning. *Quarterly Journal of Experimental Psychology, 64*(5), 1021–1040.
Evans, J. L., Saffran, J. R., & Robe-Torres, K. (2009). Statistical learning in children with specific language impairment. *Journal of Speech, Language, and Hearing Research, 52*, 321–335.
Fiser, J., & Aslin, R. N. (2001). Unsupervised statistical learning of higher-order spatial structures from visual scenes. *Psychological Science, 12*(6), 499–504.
Fiser, J., & Aslin, R. N. (2002). Statistical learning of new visual feature combinations by infants. *Proceedings of the National Academy of Sciences of the United States of America, 99*(24), 15822–15826.
Fiser, J., Berkes, P., Orbán, G., & Lengyel, M. (2010). Statistically optimal perception and learning: From behavior to neural representations. *Trends in Cognitive Sciences, 14*(3), 119–130.
Friederici, A. D., Mueller, J. L., & Oberecker, R. (2011). Precursors to natural grammar learning: Preliminary evidence from 4-month-old infants. *PLoS ONE, 6*(3), Article e17920.
Frost, R., Siegelman, N., Narkiss, A., & Afek, L. (2013). What predicts successful literacy acquisition in a second language? *Psychological Science, 24*(7), 1243–1252.
Gabay, Y., Thiessen, E. D., & Holt, L. L. (2015). Impaired statistical learning in developmental dyslexia. *Journal of Speech, Language, and Hearing Research, 58*(3), 934–945.
Gomez, R. L. (2002). Variability and detection of invariant structure. *Psychological Science, 13*(5), 431–436.
Ho, C. S. H., & Bryant, P. (1997). Development of phonological awareness of Chinese children in Hong Kong. *Journal of Psycholinguistic Research, 26*, 109–126.
Ho, C.S.-H., Yau, P.W.-Y., & Au, A. (2003). Development of orthographic knowledge and its relationship with reading and spelling among Chinese kindergarten and primary school children. In C. McBride-Chang & H.-C. Chen (Eds.), *Reading development in Chinese children* (pp. 51–71). Praeger.
Hsiao, J. H. W., & Shillcock, R. (2006). Analysis of a Chinese phonetic compound database: Implications for orthographic processing. *Journal of Psycholinguistic Research, 35*(5), 405–426.
Huang, C. R., & Chen, K. J. (1998). *Academia Sinica balanced corpus.* www.sinica.edu.tw/SinicaCorpus/Ja
Hung, Y.-H., Hung, D. L., Tzeng, O.J.-L., & Wu, D. H. (2014). Tracking the temporal dynamics of the processing of phonetic and semantic radicals in Chinese character recognition by MEG. *Journal of Neurolinguistics, 29*, 42–65.
Jost, E., Conway, C. M., Purdy, J. D., Walk, A. M., & Hendricks, M. A. (2015). Exploring the neurodevelopment of visual statistical learning using event-related brain potentials. *Brain Research, 1597*, 95–107.
Kirkham, N. Z., Slemmer, J. A., & Johnson, S. P. (2002). Visual statistical learning in infancy: Evidence for a domain general learning mechanism. *Cognition, 83*(2), B35–B42.
Kooijman, V., Hagoort, P., & Cutler, A. (2005). Electrophysiological evidence for prelinguistic infants' word recognition in continuous speech. *Cognitive Brain Research, 24*(1), 109–116.
Lee, C. Y., Tsai, J. L., Su, E. C. I., Tzeng, O. J., & Hung, D. L. (2005). Consistency, regularity, and frequency effects in naming Chinese characters. *Language and Linguistics-Taipei-, 6*(1), 75.
Maye, J., Werker, J. F., & Gerken, L. (2002). Infant sensitivity to distributional information can affect phonetic discrimination. *Cognition, 82*(3), B101–B111.
Misyak, J. B., & Christiansen, M. H. (2012). Statistical learning and language: An individual differences study. *Language Learning, 62*(1), 302–331.
Misyak, J. B., Christiansen, M. H., & Tomblin, J. B. (2010). On-line individual differences in statistical learning predict language processing. *Frontiers in Psychology, 1*, 31.

Mueller, J. L., Hahne, A., Fujii, Y., & Friederici, A. D. (2005). Native and nonnative speakers' processing of a miniature version of Japanese as revealed by ERPs. *Journal of Cognitive Neuroscience, 17*(8), 1229–1244.

Newport, E. L., & Aslin, R. N. (2004). Learning at a distance I. Statistical learning of non-adjacent dependencies. *Cognitive psychology, 48*(2), 127–162.

Pavlidou, E. V., Louise Kelly, M., & Williams, J. M. (2010). Do children with developmental dyslexia have impairments in implicit learning? *Dyslexia, 16*(2), 143–161.

Pacton, S., Perruchet, P., Fayol, M., & Cleeremans, A. (2001). Implicit learning out of the lab: The case of orthographic regularities. *Journal of Experimental Psychology: General, 130*(3), 401.

Pearce, M. T., Ruiz, M. H., Kapasi, S., Wiggins, G. A., & Bhattacharya, J. (2010). Unsupervised statistical learning underpins computational, behavioural, and neural manifestations of musical expectation. *NeuroImage, 50*(1), 302–313.

Perfors, A., & Kidd, E. (2022). The role of stimulus-specific perceptual fluency in statistical learning. *Cognitive Science, 46*(2), Article e13100.

Renaldi, A., Avraham, E., Belson, R., Siegelman, N., Frost, R., & Wu, D.H. *Superior performance of logographic than alphabetic readers in the visual spatial positional statistical learning test.*

Renaldi, A., & Wu, D.H. *Statistical learning contributes to acquisition of Chinese character recognition: Evidence from a novel positional statistical learning paradigm.*

Renaldi, A., & Wu, D.H. *Neurophysiological components supporting positional statistical learning on different orthography readers: An ERP Study.*

Rothe, J., Cornell, S., Ise, E., & Schulte-Körne, G. (2015). A comparison of orthographic processing in children with and without reading and spelling disorder in a regular orthography. *Reading and Writing, 28*(9), 1307–1332.

Saffran, J. R. (2002). Constraints on statistical language learning. *Journal of Memory and Language, 47*, 172–196.

Saffran, J. R., Aslin, R. N., & Newport, E. L. (1996). Statistical learning by 8-month-old infants. *Science, 274*(5294), 1926–1928.

Sanders, L. D., Newport, E. L., & Neville, H. J. (2002). Segmenting nonsense: An event-related potential index of perceived onsets in continuous speech. *Nature Neuroscience, 5*(7), 700–703.

Seidenberg, M. S., & McClelland, J. L. (1989). A distributed, developmental model of word recognition and naming. *Psychological Review, 96*(4), 523.

Siegelman, N., Bogaerts, L., Elazar, A., Arciuli, J., & Frost, R. (2018). Linguistic entrenchment: Prior knowledge impacts statistical learning performance. *Cognition, 177*, 198–213.

Siegelman, N., & Frost, R. (2015). Statistical learning as an individual ability: Theoretical perspectives and empirical evidence. *Journal of Memory and Language, 81*, 105–120.

Sigurdardottir, H. M., Danielsdottir, H. B., Gudmundsdottir, M., Hjartarson, K. H., Thorarinsdottir, E. A., & Kristjánsson, Á. (2017). Problems with visual statistical learning in developmental dyslexia. *Scientific Reports, 7*(1), 606.

Tabullo, Á., Sevilla, Y., Segura, E., Zanutto, S., & Wainselboim, A. (2013). An ERP study of structural anomalies in native and semantic free artificial grammar: Evidence for shared processing mechanisms. *Brain Research, 1527*, 149–160.

Teinonen, T., Fellman, V., Näätänen, R., Alku, P., & Huotilainen, M. (2009). Statistical language learning in neonates revealed by event-related brain potentials. *BMC Neuroscience, 10*(1), 21.

Thiessen, E. D. (2010). Effects of visual information on adults' and infants' auditory statistical learning. *Cognitive Science, 34*(6), 1093–1106.

Thiessen, E. D., & Erickson, L. C. (2013). Beyond word segmentation: A two-process account of statistical learning. *Current Directions in Psychological Science, 22*(3), 239–243.

Tong, X., Leung, W. W. S., & Tong, X. (2019). Visual statistical learning and orthographic awareness in Chinese children with and without developmental dyslexia. *Research in Developmental Disabilities, 92*, Article 103443.

Tsai, P. S., Tzeng, O. J. L., Hung, D. L., & Wu, D. H. (2013). Using magnetoencephalography to investigate processing of negative polarity items in Mandarin Chinese. *Journal of Neurolinguistics, 26*(2), 258–270.

Tzeng, Y. L., & Lee, C. Y. (2012). The developmental trajectories of orthography-to-phonology mapping consistency in learning to read Chinese. *Contemporary Educational Research Quarterly, 20*(4), 45–84.

Wu, D. H., & Bulut, T. (2020). The contribution of statistical learning to language and literacy acquisition. In *Psychology of Learning and Motivation* (Vol. 72, pp. 283–318). Academic Press.

Xu, K., Duann, J. R., Hung, D. L., & Wu, D. H. (2019). Preference for object relative clauses in Chinese sentence comprehension: Evidence from online self-paced reading time. *Frontiers in Psychology, 10*, Article 476094.

Yang, J., McCandliss, B. D., Shu, H., & Zevin, J. D. (2009). Simulating language-specific and language-general effects in a statistical learning model of Chinese reading. *Journal of Memory and Language, 61*(2), 238–257.

Yu, A., Chen, M. S., Cherodath, S., Hung, D. L., Tzeng, O. J., & Wu, D. H. (2019). Neuroimaging evidence for sensitivity to orthography-to-phonology conversion in native readers and foreign learners of Chinese. *Journal of Neurolinguistics, 50*, 53–70.

Ziegler, J. C., & Goswami, U. (2005). Reading acquisition, developmental dyslexia, and skilled reading across languages: A psycholinguistic grain size theory. *Psychological Bulletin, 131*(1), 3.

van Zuijen, T. L., Simoens, V. L., Paavilainen, P., Näätänen, R., & Tervaniemi, M. (2006). Implicit, intuitive, and explicit knowledge of abstract regularities in a sound sequence: An event-related brain potential study. *Journal of Cognitive Neuroscience, 18*(8), 1292–1303.

Part III
Reading Revealed and Facilitated by Advanced Technology

Chapter 8
Sophisticated Neuroimaging Methods to Study the Brain Circuit for Language Comprehension

Jeng-Ren Duann and Kunyu Xu

Abstract Traditional fMRI data analysis relies on pre-defined BOLD reference functions derived from task paradigms, potentially overlooking transient brain processes linked to task performance. Additionally, active brain areas identified through general linear model (GLM) analysis exhibit highly correlated time courses, limiting effective connectivity analysis. To address these issues, we applied independent component analysis (ICA) to extract independent BOLD processes and identify task-related components. This approach preserves the time courses of selected independent BOLD components, enabling connectivity analysis without the constraints of high correlation inherent in GLM-based methods. More importantly, ICA allows for single-trial segmentation of BOLD time courses, facilitating single-trial analysis. By leveraging this methodology, we assessed functional connectivity at the single-trial level and examined its correlation with the difficulty of individual Chinese relative clause sentences. This novel approach enhances our ability to investigate dynamic neural processes and their relationship to cognitive complexity, providing deeper insights into sentence comprehension mechanisms.

Keywords Functional Magnetic Resonance Imaging (fMRI) · Independent Component Analysis (ICA) · Surface-based Spatial Normalization · Deep Neural Network (DNN) · Brain connectivity

---

J.-R. Duann (✉)
Institute of Education, National Yang Ming Chiao Tung University, Hsinchu, Taiwan
e-mail: duann@nycu.edu.tw

Institute for Neural Computation, University of California San Diego, La Jolla, CA, USA

K. Xu
Institute of Modern Languages and Linguistics, Fudan University, Shanghai, China

D. H. Wu (ed.), *Cognitive and Neural Foundations of Chinese Reading*, Chinese Language Learning Sciences, https://doi.org/10.1007/978-981-96-6669-0_8

8.1 Introduction

The examination of how the brain responds to processes of varying difficulty levels has long captivated the interest of neuroscientists. However, isolating issues characterized by clear-cut differences in difficulty, free from confounding factors, has proven a challenging endeavor. Recently, the comprehension of relative clauses has emerged as a particularly compelling subject of investigation within the field of functional neuroimaging. This is due to the fact that, through strategic manipulation of word and phrase placement, relative clauses can be designed to induce differing levels of comprehension difficulty. Notably, a series of functional neuroimaging studies has shed light on the phenomenon of SRC-advantage, where English subject-extracted relative clauses (SRC) are found to be relatively easier to comprehend compared to English object-extracted relative clauses (ORC), as reported in studies such as King and Just (1991). Furthermore, relative clauses are not unique to English but are a linguistic feature found in multiple languages. This universality affords researchers the opportunity to explore the SRC advantage across diverse linguistic contexts, thereby probing the potential universality of this phenomenon.

In fact, the experimental protocol used to discern the difficulty levels of English relative clauses, primarily based on eye-tracking experiments, has also been applied to assess the complexity of Chinese relative clauses. It is noteworthy that while eye-tracking experiments have consistently provided valuable insights into the advantage of subject-extracted relative clauses (SRC) in English relative clause comprehension, it is important to acknowledge that this single modality tool may not yield equivalent insights when applied to Chinese relative clause comprehension. Indeed, the use of eye tracking to investigate processing times in the comprehension of Chinese relative clauses has generated somewhat contradictory findings. Some studies have suggested prolonged processing times for SRC sentences (e.g., Hsaio & Gibson, 2003), while others have reported extended processing times for ORC sentences (e.g., Jager et al., 2015). These variations underscore the need for a multimodal approach to comprehensively elucidate the intricate dynamics of relative clause processing in the Chinese language.

In light of these disparate findings, our approach was to employ a multimodal neuroimaging framework, which integrates functional magnetic resonance imaging (fMRI) blood oxygenation-level dependent (BOLD) signal and eye tracking data. We began by adhering to the traditional experimental paradigm to assess the processing times associated with different types of Chinese relative clauses. Given the historical variability in results from eye-tracking studies, we embarked on an fMRI experiment to augment our understanding of the neural processes at play, striving for a convergence of evidence through the harmonization of fMRI and eye-tracking paradigms.

This approach enables us to not only confirm the behavioral aspects of sentence comprehension but also gain deeper insights into the underlying neural mechanisms, transcending the limitations of a single modality tool. Our aim is to elucidate the complex interplay between linguistic structures and cognitive processes, drawing from the strengths of both fMRI and eye tracking to provide a more comprehensive understanding of Chinese relative clause comprehension.

8.2 Complexity Level of SRC and ORC Using Eye Tracker

To embark on a comprehensive exploration of the intricate neural mechanisms underpinning Chinese relative clause comprehension, we meticulously crafted a corpus of 64 pairs of Chinese SRC and ORC short sentences, resulting in a total of 128 sentences for our investigation. Crucially, these sentences were meticulously designed to maintain equality in terms of overall word length between the two conditions, adhering to a consistent structural framework. Each sentence was thoughtfully divided into six frames, with each frame accommodating one to three characters, aligning seamlessly with the behavioral self-paced reading paradigm we had employed in a prior study (Xu et al., 2019a).

Furthermore, all the experimental sentences were characterized by the presence of two animate nouns or noun phrases, each of which had an equal likelihood of serving as either the subject or object of the verb, rendering the sentences semantically reversible (see Example A and B in Fig. 8.1). To ensure the suitability of these sentence structures for our study, a pilot investigation entailing plausibility ratings was conducted with a separate group of 20 participants who were not part of the fMRI study. The results from this pilot study revealed no statistically significant differences in the perceived naturalness or plausibility between these two sentence types, thereby affirming their appropriateness for our research objectives ($t1\ (19) = 1, p = 0.330, t2\ (1\,2\,6) = 0.982, p = 0.328$). This meticulous construction of sentence stimuli provided a solid foundation for our subsequent investigations into the neural underpinnings of Chinese relative clause processing. The example of the stimuli of the SRC and ORC sentences is shown in Fig. 8.1.

To assess the temporal dynamics of SRC and ORC comprehension through eye-tracking methodologies, a cohort of 46 native Chinese-speaking students (comprising 26 females) with ages ranging from 20 to 28 years was recruited from National Central University in Taoyuan, Taiwan. All recruited participants attested to their right-handedness and exhibited normal or corrected-to-normal vision, ensuring the homogeneity of the study sample. It is noteworthy that every participant provided explicit, written informed consent, aligning with the ethical guidelines enshrined in the Declaration of Helsinki, which prioritize the welfare and rights of research subjects. Importantly, the experimental protocol received the full and official approval of the Research Ethics Committee at National Taiwan University, reinforcing the commitment to ethical research practices and human subject protection.

The eye-tracking experiment was executed, commencing with a crucial calibration phase to ensure precision and accuracy in gaze tracking. To maintain the integrity of gaze tracking, the eye tracker was subject to periodic recalibration, conducted at the conclusion of each block of sentences or as deemed necessary by the experimenter to counteract any potential loss of fixation stability.

a. Subject-extracted relative clause sentence (SRC)					
AOI1	AOI2	AOI3	AOI4	AOI5	AOI6
攻擊	張三	的	司機	違反了	規定
gōngjī	Zhāngsān	de	sījī	wěifǎnle	guīdìng
attack	Zhangsan	de	driver	violate	rule
'The driver who attacked Zhangsan has violated the rules.'					
b. Object-extracted relative clause sentence (ORC)					
AOI1	AOI2	AOI3	AOI4	AOI5	AOI6
張三	攻擊	的	司機	違反了	規定
Zhāngsān	gōngjī	de	sījī	wěifǎnle	guīdìng
Zhangsan	attack	de	driver	violate	rule
'The driver who Zhangsan attacked has violated the rules.'					

Fig. 8.1 Example of the stimuli: (a) Subject-extracted relative clause (SRC) sentence, (b) Object-extracted relative clause (ORC) sentence (reproduced from Xu et al., 2020a, 2020b)

Throughout the experimental proceedings, all sentences were presented in the DFKai-SB font and were individually randomized in presentation order following six initial practice trials. Participants were provided with explicit instructions to meticulously read each sentence at their own pace, subsequently signaling their comprehension and completion of the sentence by pressing a designated button. After reading a sentence, participants were also tasked with answering a true/false comprehension question, designed with the structure of "who did what to whom," a measure to gauge their comprehension of the sentence's meaning.

The experimental protocol consisted of two sessions, each comprising 64 sentences. A scheduled break between these sessions afforded participants a brief intermission. In its entirety, including the calibration, practice trials, the two experimental blocks, and the intermission, the experiment was designed to conclude in approximately 45 min, ensuring that the participants' cognitive engagement and alertness remained consistently maintained throughout the duration of the study.

The meticulous analysis of the eye-tracking data encompassed a comprehensive examination of fixation locations, durations, and timings. To ensure data integrity, individual fixations exceeding 800 ms or falling below 80 ms were automatically removed. A multifaceted array of eye-movement metrics, including First Fixation Duration, Gaze Duration, Go-Past Time, Total Fixation Time, and Number of Fixations in the six predefined Areas of Interest (AOIs) was scrutinized.

To ensure the robustness and statistical reliability of the observed reading time effects, the data underwent rigorous scrutiny via a mixed-effects model. The model featured sentence type as the fixed factor, with subjects and items incorporated as random intercepts and slopes—a statistical approach renowned for its ability to address nested data structures and random variability (Baayen et al., 2008). The model was constructed and assessed employing the lme4 package and the lmerTest package within the R software environment (R Core Team, 2011).

The results of the statistical analyses, conducted through a linear mixed-effects model to investigate latency measures, revealed compelling insights into the influence of clause type (ORC and SRC) on various eye-tracking metrics at specific Areas of Interest (AOIs). Notably, the impact of clause type was found to be highly significant at AOI1 (estimate = −0.29, SE = 0.11, t = −2.736, p = 0.008), AOI2 (estimate = −0.39, SE = 0.14, t = −2.748, p = 0.008), and AOI4 (estimate = −0.19, SE = 0.08, t = −2.29, p = 0.030).

This compelling statistical evidence pointed to the heightened cognitive demands associated with reading SRC sentences, reflected in a greater number of fixations, particularly within the embedded clause (i.e., AOI1 and AOI2) and the head noun (i.e., AOI4) regions. Specifically, the reading of SRC sentences necessitated an increased number of fixations at AOI1 (2.40 vs. 2.23), AOI2 (3.55 vs. 3.13), and AOI4 (2.88 vs. 2.73) compared to the reading of ORC sentences.

Moreover, the analysis also unveiled significantly shorter total fixation times at AOI1 (estimate = −85.41, SE = 27.09, t = −3.152, p = 0.002) and AOI4 (estimate = −44.42, SE = 19.19, t = −2.314, p = 0.028) during the comprehension of ORC sentences (AOI1: 539.61 ms; AOI2: 700.00 ms; AOI4: 571.37 ms) compared to SRC sentences (AOI1: 596.44 ms; AOI2: 747.02 ms; AOI4: 608.61 ms). Although the total fixation time on AOI2 was numerically shorter for ORC sentences (estimate = −39.93, SE = 29.35, t = −1.61, p = 0.18), this difference did not reach statistical significance.

In contrast, the manipulation of clause type (AOI1, AOI2, or AOI4) exhibited no discernible effects on other eye-movement measures, including go-past time, gaze duration, and first fixation time. Additionally, no statistically significant disparities in eye-movement metrics were detected for the other AOIs (i.e., AOI3, AOI5, and AOI6, ps > 0.115).

As a result, the quantification of sentence complexity was achieved by summing the number of fixations or the total fixation time within the embedded clause (AOI1 and/or AOI2) and the head noun (AOI4) regions, which were demonstrated to be indicative of a significant differentiation between SRC and ORC sentences. This comprehensive analysis provides invaluable insights into the cognitive processes underpinning the comprehension of Chinese relative clauses and indicates an ORC advantage in Chinese relative clauses.

8.3 Conventional fMRI Analysis to Reveal Different Brain Activities in Response to Reading SRC and ORC

To offer a comprehensive perspective through multiple functional neuroimaging modalities, we complemented our investigation with an fMRI experiment employing the same set of relative clause materials used in the eye-tracking study. The 64 experimental sentence pairs encompassing both Chinese SRCs and ORCs were systematically divided into two distinct lists, each comprising 32 SRC sentences and 32 ORC sentences. Crucially, SRC and ORC sentences within the same pair were distributed across separate lists. Half of the participants were presented with one list, while the remaining participants received the alternate list, ensuring that no participant encountered both SRC and ORC sentences within a pair during the course of the experiment.

Each of the 64 sentences within a list was further equitably distributed across four sub-lists, one for each of the four fMRI runs. The sequence of sentences within these sub-lists was completely randomized across participants, minimizing order-related biases. Following the presentation of a complete target sentence, participants were promptly tasked with a true/false judgment in response to a probe question, accomplished by pressing either a left or right-handed button using their left or right thumb within a 4-s window. The temporal interval between the target sentence and the subsequent probe question was purposefully jittered, fluctuating between 2 and 4s. Notably, half of the probe questions required a "true" response, while the other half necessitated a "false" response for correct answers. Immediate feedback concerning each participant's responses was provided promptly following their response, thereby promoting an environment conducive to rigorous data collection and analysis.

In our pursuit to distinguish brain areas primarily involved in the comprehension of RC sentences from those associated with fundamental visual stimulus processing, we introduced a visual orientation task as a control condition. Specifically, participants were presented with visual stimuli featuring two rows of five arrows, with orientations either matching or differing (e.g., "> > > > >" and "< < > < <"). These rows were positioned one above the other, and participants were tasked with a same/different orientation judgment, carried out by pressing the corresponding button using their left or right thumb. In this baseline condition, each trial commenced with a 1000-ms fixation period, followed by the display of the dual rows of arrows, where participants had a 3-s window to formulate their judgment. This strategic inclusion of a control task allowed us to scrutinize and isolate the neural substrates primarily implicated in RC sentence comprehension.

Nineteen participants (9 females, aged 19–29 years, mean age = 23 years, standard deviation = 2.3 years), all right-handed native Chinese speakers, were remunerated for their participation in the study. The screening process initially involved 21 volunteers; however, 2 were subsequently excluded due to their low accuracy in sentence comprehension questions (71.88% and 68.75%). All the individuals who continued to participate exhibited normal or corrected-to-normal vision and had no history of neurological or psychiatric disorders. The study adhered to the ethical standards established by the Research Ethics Committee of National Taiwan University (NTU), Taipei, Taiwan. It was conducted in strict accordance with the guidelines outlined in the Social and Behavioral Research Ethical Principles and Regulations of NTU, with every participant providing signed written informed consent prior to their involvement.

Magnetic resonance imaging (MRI) was conducted using a 3 T scanner (MAGNETOM Skyra, Siemens, Erlangen, Germany) equipped with a 64-channel whole-head coil at National Cheng-Chi University in Taipei, Taiwan. Functional data, obtained through blood oxygenation level-dependent (BOLD) signals, were acquired using a T2*-weighted echo planar imaging (EPI) sequence. This sequence featured a slice thickness of 3.4 mm with no interslice gap, an in-plane resolution of 3.4375 × 3.4375 mm, and a TR/TE/flip angle of 2000 ms/70 ms/77°. The field-of-view covered 220 × 220 mm, and the acquisition matrix was set at 64 × 64. A total of 34 oblique-axial slices, oriented parallel to the anterior commissure-posterior commissure (AC-PC) line, were captured to encompass the entire brain. The initial five volumes from each functional run were discarded to ensure signal equilibrium, leaving one hundred and eighty image volumes for analysis in each of the four runs. In addition to functional data, high-resolution anatomical images of each participant's brain were obtained through a T1-weighted sequence (TR = 2530 ms/ TE = 3.3 ms/flip angle = 7°, bandwidth = 200 Hz/pixel). This sequence featured 192 sagittal slices to comprehensively cover the entire brain, achieving an isotropic spatial resolution of $1 \times 1 \times 1\text{mm}^3$, facilitating detailed localization of fMRI activity.

In this part of fMRI data analysis, we used a traditional general linear model-based approach with the Statistical Parametric Mapping package (SPM8, https://www.fil.ion.ucl.ac.uk/spm/). The image data with error responses were excluded from the analysis. Images of each participant were first adjusted for slice timing so that each slice's time series was temporally aligned after correcting the shifts between the slice acquisitions in a volume. Then, the image volume series was realigned to the middle image volume of each run to correct for head movements. A mean functional image volume was created for each participant and each run from the realigned image volumes and spatially normalized to an MNI (Montreal Neurological Institute) EPI template using a nonlinear affine transformation. The normalization parameters determined from the mean functional volume were then applied to the corresponding functional image volumes for each participant. Finally, the images were smoothed with a Gaussian kernel of 8 mm full-width at half-maximum.

An analytical statistical design was constructed for each participant in this study. In the first-level analysis, two essential contrasts were generated for each participant, effectively comparing the sentence comprehension conditions against the visual orientation (VO) task. These contrasts included SRC vs. VO and ORC vs. VO. Subsequently, the 'con' or contrast images obtained from the first-level analysis were harnessed for the second level group statistics through a one-sample T-test. To maintain rigorous control over Type I errors, we applied family-wise error (FWE) correction to account for multiple comparison errors.

At the single-subject level, brain activation maps were thresholded at $p < 0.05$, with corrections for multiple comparisons using FWE. To facilitate the identification of target brain areas, we employed the xjView toolbox (http://www.alivelearn.net/xjview), setting a minimum cluster size threshold of > 10 voxels. Additionally, we utilized MarsBaR (http://marsbar.sourceforge.net/) to extract effect sizes representing activity changes in the identified brain regions for each participant. This methodological approach ensured a robust and comprehensive analysis of the neural responses to the diverse sentence comprehension conditions.

Intriguingly, beyond the insights gleaned from observing the activation patterns generated by regression models, the exploration of temporal precedence information within fMRI data can be greatly enriched through the application of Granger Causality (GC) analysis. GC, initially introduced by Granger in 1969, stands as a powerful and indispensable tool in the domain of neuroscience. It allows us to identify specific brain regions that potentially serve as sources or recipients of directed influences in relation to other brain areas based on fMRI time-series data. This methodology has evolved into a cornerstone of research, extensively embraced by numerous scholars in the field. Notably, its application has been instrumental in studies conducted by esteemed researchers such as Deshpande et al. (2009), Duann et al. (2009), Goebela et al. (2003), Roebroeck et al. (2005), and Sato et al. (2009). The utilization of GC analysis in conjunction with other analytical techniques empowers neuroscientists to uncover the intricate network of causal relationships that underlie brain function and connectivity.

In technical terms, GC provides a means of quantifying the directional influence between different brain areas by examining the delayed temporal information from these regions. For example, consider two distinct time courses, denoted as x[n] and y[n]. The GC estimation process entails calculating the variance, denoted as ε, of the time series data y[n] accounted for solely by past values of y[n] within an autoregressive (AR) model. Subsequently, we incorporate past values of the complementary time series data x[n] to further explain the signal y[n], resulting in the residual variance ε'. When ε', the residual variance of the regression model including x[n], is found to be significantly smaller than ε, the signal x[n] is deemed to Granger-cause the signal y[n], based on the definition of GC as described by Roebroeck et al. (2005).

Similarly, the reciprocal relationship can be examined by computing the residual variance of x[n] using an AR model, both with and without the inclusion of past values of y[n], to ascertain whether y[n] significantly Granger-causes the signal x[n]. Notably, it is imperative to recognize that the determination that x[n] Granger-causes y[n] does not inherently imply the reverse causal relationship. Hence, the directionality of this causality estimation underscores its unique attribute, aptly referred to as effective connectivity. This analytical approach serves as a powerful instrument in elucidating the intricate network of directional influences within the brain.

Following the initial contrasts between each sentence type and the visual baseline, a subsequent direct contrast analysis was conducted between the two specific conditions, namely (SRC - VO) and (ORC - VO). This analytical approach aimed to elucidate the distinct neural activation patterns associated with sentence comprehension. The findings unveiled a notably enhanced activation within the left inferior frontal gyrus (LIFG) and the left superior temporal gyrus (LSTG) in response to SRC sentences compared to ORC sentences. These significant differences were observed at a threshold of $p < 0.001$, as depicted in Fig. 8.2. In contrast, the analysis of (ORC - VO) vs. (SRC - VO) did not yield any significant results, signifying the absence of regions exhibiting higher activation during ORC sentence comprehension than during SRC sentence comprehension. These results underscore the specific engagement of the LIFG and LSTG in processing SRC sentences, reflecting the complexity and unique demands associated with this sentence structure.

Furthermore, the results obtained from the GC analysis provided valuable insights into the directed functional interactions between brain regions during the comprehension of different sentence types. Notably, when processing the more challenging SRC sentences, as determined by the whole-brain analysis, the LIFG exhibited a

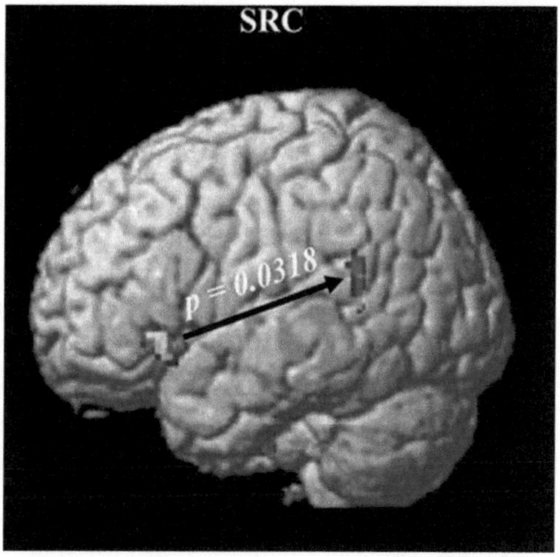

Fig. 8.2 The active brain areas, namely, the LIFG and the LSTG and their connectivity derived by the traditional GLM-based fMRI data analysis method and Granger causality analysis

statistically significant Granger causality effect on the LSTG at the group level. This was confirmed through a rigorous binomial test ($p = 0.0318$), as illustrated in Fig. 8.2. Conversely, the Granger causality effect in the same direction when reading ORC sentences failed to reach statistical significance ($p = 0.9682$).

Conversely, the Granger causality from the LSTG to the LIFG did not reach statistical significance in either sentence condition (SRC: $p = 0.6762$; ORC: $p = 0.6762$), and the direct comparison of the Granger causality effects between the two conditions did not reveal a significant difference ($Z = 1.288$, $p = 0.210$).

Additionally, the Wilcoxon Signed Ranks test provided further confirmation of these findings. In the SRC condition, the strength of connectivity from the LIFG to the LSTG was significantly stronger than in the opposite direction ($Z = 2.334$, $p = 0.018$). Furthermore, the connectivity strength from the LIFG to the LSTG in the SRC condition was also significantly greater than that observed in the ORC condition ($Z = 2.575$, $p = 0.008$). These results shed light on the intricate dynamics of information flow within the language processing network, emphasizing the specific role of LIFG-to-LSTG connectivity in handling the increased demands of SRC sentence comprehension.

8.4 Limitations of the Traditional fMRI Data Analysis Results for Further Granger Causality Analysis

One of the significant challenges inherent in the computation of connectivity based on the time courses extracted through conventional fMRI data analysis lies in the nature of the functioning brain regions that withstand statistical scrutiny employing a general linear model (GLM). These regions are fundamentally reliant upon the reference function(s) crafted through the convolution of the task time course with the canonical hemodynamic response function (HRF). Consequently, these active brain regions inherently exhibit a high degree of shared, and therefore correlated, BOLD time courses from the outset. This inherent correlation poses a fundamental dilemma, as any subsequent derivation of connectivity indices from the time courses of these activated regions, irrespective of the methodology employed, is susceptible to a marked overestimation. This quandary not only casts doubt on the validity of fMRI-based connectivity measures in a general sense but also compounds the challenge of interpreting the resultant connectivity metrics. For example, one may need to partial out the correlation between the time courses before the connectivity information can be derived. Consequently, an fMRI data analysis that does not harness the temporal information inherent in the fMRI data becomes highly desirable when subsequent connectivity metrics are to be derived from the identified functioning brain regions established through activation analysis.

8.5 Independent Component Analysis

Independent Component Analysis (ICA) was initially conceived as a solution to the challenging problem akin to a "cocktail party scenario": Can we, armed only with recorded outputs and devoid of knowledge regarding the sources of sound or the mixing process, mathematically discern the identity of sound sources and the intricacies of their amalgamation during recording? In the realm of fMRI data analysis, the recorded output represents a complex amalgamation of various sources, encompassing the desired cerebral BOLD activations, regrettably accompanied by undesired elements such as physiological pulsations (e.g., cardiac and respiratory rhythms) and machine-induced artifacts.

When ICA is applied to these intricate fMRI mixtures, it manifests as an apt tool for the blind disentanglement of the original fMRI recordings into independent components corresponding to the brain's BOLD responses, as well as a collection of artifactual components. In effect, during this decomposition process, ICA seeks to accomplish two pivotal objectives: firstly, to disentangle these independent processes from the raw data and, secondly, to organize the voxels with highly correlated separated component time courses into spatially discrete regions (thus, spatial independence), thereby generating the component maps. Figure 8.3 serves as an exemplar of an independent BOLD component resulting from an ICA decomposition. The lower trace within the figure delineates the BOLD time course for the visual component, uncontaminated by extraneous processes. The superimposed maps above vividly portray the spatial extent of this visual component across the entirety of the brain, with color hues elucidating the magnitude of each voxel's correspondence with the component's time course. Furthermore, it is worth noting that the raw time course of each voxel may be influenced by multiple independent components, each exhibiting varying degrees of influence, as visually represented by the color gradients within

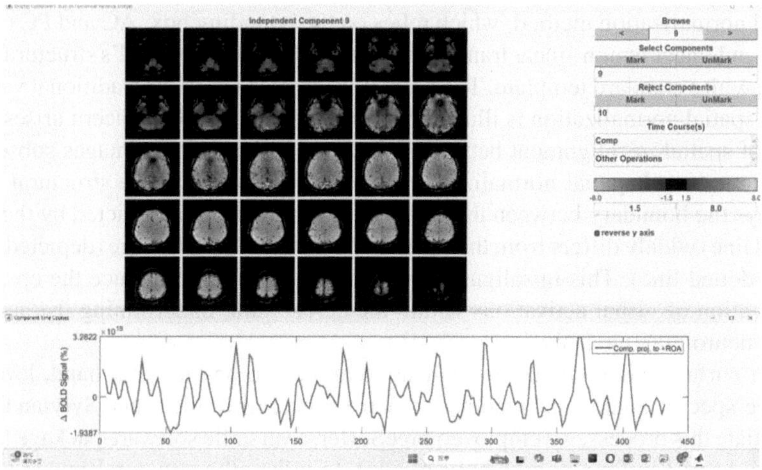

Fig. 8.3 The outcomes of one selected independent BOLD component resulted from an ICA decomposition (component browser of the FMRLAB Matlab toolbox)

the component maps. This distributed influence can, in turn, lead to spatial overlap among these component maps.

Adopting this methodological approach allows us to leverage the time courses of distinct independent components for the computation of connectivity metrics, as these time courses have not been employed in the selection of active brain regions through statistical comparisons. However, it is important to note that the time course of a single independent component is typically shared by the prominently active voxels within the same component, resulting in notably high correlations among them. Consequently, the time courses of active voxels within the same independent components are unsuitable for deriving connectivity measures between the active voxels within the same independent components.

In contrast, when considering different components, the time courses exhibit significant utility for computing connectivity between various brain regions delineated by separate independent BOLD components. This distinction is pivotal in establishing robust and meaningful connectivity assessments across the brain.

8.6 Selecting the Independent BOLD Components of the Reading Brain Circuit

In order to systematically identify spatially congruent independent BOLD components associated with the intricate language comprehension circuit, encompassing regions like the inferior frontal gyrus and the superior temporal gyrus, across diverse participants, we have implemented an advanced surface-based spatial normalization technique. This approach distinguishes itself from the conventional volume-based spatial normalization method, which relies on the bounding box, AC and PC coordinates, and linear or non-linear transformations to align an individual's structural brain image with a standard template. The problem associated with the traditional volume-based spatial normalization is illustrated in Fig. 8.4. A notable concern arises in the form of spatial misalignment between functional and structural images subsequent to volume-based spatial normalization. Evidently, on the left, the structural image portrays the boundary between the cerebrum and cerebellum (depicted by the black dotted line) widely differs from the boundary in the functional image (depicted by the white dotted line). This misalignment carries the potential to induce the erroneous localization of visual activations within the cerebellum, undermining the accuracy of the neuroimaging analysis.

Our surface-based spatial normalization approach, on the other hand, leverages surface-specific anatomical landmarks such as the central sulcus and Sylvian fissure. To initiate this process, we employed FreeSurfer, a versatile software package (accessible at https://surfer.nmr.mgh.harvard.edu), to inflate the structural image of each participant. Subsequently, the inflated individual brain was meticulously mapped to the inflated cortical surface of the standardized brain template, utilizing a spherical transformation methodology designed to preserve the essential cortical landmarks.

By following this procedure, we are able to overlay activation maps of independent BOLD components on the inflated brain template, enabling the precise selection of spatially congruent components across participants. This surface-based approach not only enhances the accuracy of our analyses but also provides a robust foundation for investigating the language comprehension network within the human brain.

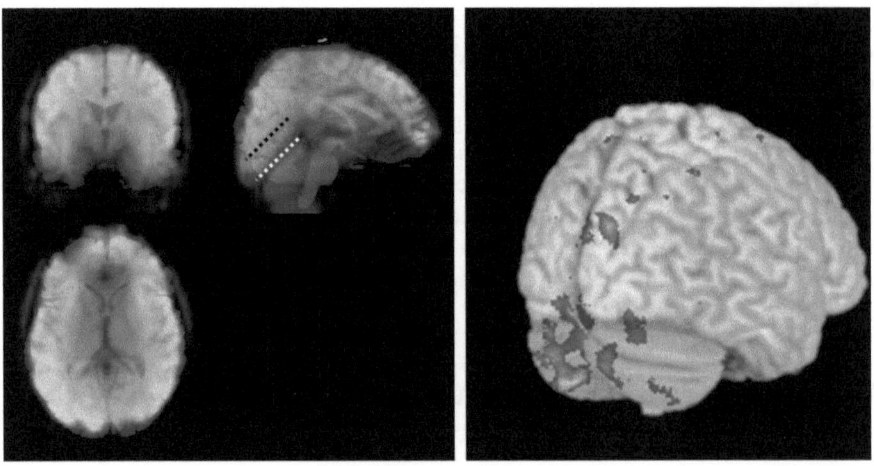

Fig. 8.4 The misalignment between the functional image and structural image after volume-based spatial normalization (white dotted line: the boundary between the cerebrum and the cerebellum in the functional image; black dotted line: the boundary between the cerebrum and the cerebellum in the structural image)

Furthermore, to find the same activated brain areas or circuits, thus, the equivalent components, from different participants, the overlap ratio measure (ORM) was computed according to Eq. (8.1).

$$\mathrm{ORM} = \mathrm{N}(\mathrm{A} \cap \mathrm{B})/\mathrm{N}(\mathrm{A} \cup \mathrm{B}) \tag{8.1}$$

Within this framework, we introduced variables A and B to represent the active voxels within the selected equivalent component of participant A and participant B, respectively. The function N(•) quantified the number of active voxels in the component map. Subsequently, we computed $A \cap B$ and $A \cup B$, signifying the intersection and union of the activation maps from the two components involved in the analysis.

In the pursuit of matching equivalent components across participants, we employed the ORM to evaluate all possible component pairs. Subsequently, we utilized a Hungarian sorting algorithm to optimize the cumulative sum of the ORMs, thus selecting the most fitting equivalent component pairs. This process resulted in the formation of an equivalent component cluster, encompassing contributions from all participants.

However, in certain cases where conflicts in component selection arose, a diligent manual inspection was conducted. This entailed a thorough examination of activation maps, both on the inflated cortical surface and in the original 2-D overlay displays. Ultimately, our selection criteria for components demonstrating brain activation in the LIFG and the LSTG relied on maximizing the number of active voxels within these regions, as observed within the inflated spherical cortical surface of the standard brain template. This comprehensive approach ensures the rigorous selection of components that align with our research objectives.

8.7 Revealing the Connectivity Between the LIFG and LSTG Independent BOLD Components Using Granger Causality

Subsequent to the identification of equivalent independent components of the LIFG and the LSTG across all participants, a meticulous analysis to compute the connectivity between these two brain areas was conducted. The BOLD time courses for each respective component within each participant were epoched based on conditions involving the SRC and ORC comprehension. These epoched time courses were then concatenated to generate the conditional time courses. This critical step allowed for a comprehensive exploration of the dynamics within the LIFG component and the LSTG component through Granger causality (GC) analysis.

Within the GC analysis, residual variance was calculated for each time course, employing a bivariate AutoRegressive (AR) model, with the optimal number of lags established using the Bayesian information criterion (BIC) for each specific time course. The evaluation of directional connectivity involved the computation of F values, enabling a comparative assessment of residual variance with and without the inclusion of the complementary time course in the bivariate AR model. At a significance level of $p < 0.05$, the directionality of connectivity between these regions was established.

To consolidate the group findings, a binomial test was subsequently employed, maintaining a significance level of $p < 0.05$, to summarize the directional connectivity between the LIFG component and the LSTG component under each condition, as described by Duann et al. in 2009. Notably, it is essential to emphasize that if both the LIFG and the LSTG happened to be situated within a single component simultaneously, it would imply that the time courses from these two brain regions exhibited a notably high degree of correlation. In essence, the GC analysis of two highly correlated time courses inherently leans toward establishing a bidirectional connectivity, a crucial insight into the intricacies of the neural network under investigation.

The Granger causality (GC) analysis conducted across all participants, furnishes insightful revelations regarding the dynamics within the LIFG and the LSTG during the comprehension of ORC and SRC sentences, as depicted in Fig. 8.5.

In the context of ORC sentences, the GC results reveal that the LIFG significantly Granger-causes the LSTG ($p = 0.0096$, confirmed through a binomial test at the group level), indicating a unidirectional influence in this scenario. In contrast, the causation in the opposite direction, namely from the LSTG to the LIFG, was not observed to be significant ($p = 0.3238$, as confirmed by the binomial test at the group level).

Conversely, during the processing of SRC sentences, the GC analysis unveils a more intricate picture. Alongside the significant connectivity observed from the LIFG to the LSTG ($p = 0.0022$, as verified by a binomial test at the group level), a noteworthy finding emerges. The connectivity from the LSTG to the LIFG is also deemed significant ($p = 0.0096$, confirmed through a binomial test at the group level). This dual-directional influence highlights the presence of reciprocal connections between

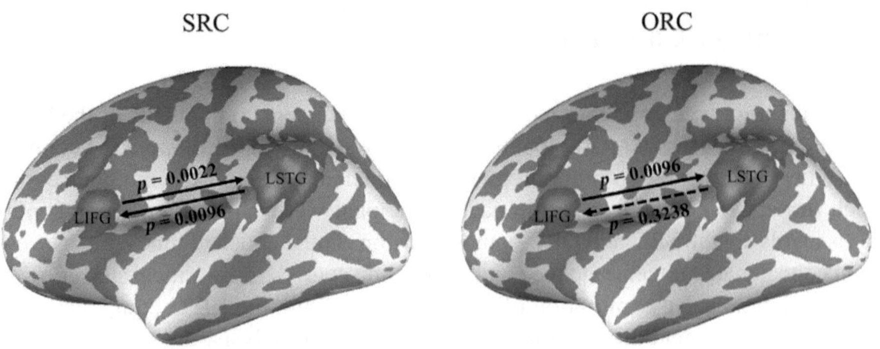

Fig. 8.5 The Granger causality between the brain areas of LIFG and LSTG under the comprehension of SRC (left) and ORC sentences (reproduced from Xu et al., 2020b)

these pivotal regions, the LIFG, and the LSTG, providing a deeper understanding of the intricate neural network underpinning language comprehension.

8.8 Computing Trial-By-Trial GC Connectivity and Correlating It with the Complexity Level of Each Sentence?

Drawing from insights proposed by Brovelli in 2012, a rigorous approach involving single-trial Granger causality (GC) analysis, coupled with robust statistical inference, was applied. In their study, this methodology was initially validated using synthetic data, comprising a two-node network model characterized by two autoregressive processes, and further extended to neurophysiological data originating from concurrent local field potentials derived from the dorsal premotor and lateral prefrontal cortex.

The findings from these investigations provided compelling evidence to support the feasibility and reliability of computing directional connectivity between bivariate signals on a trial-by-trial basis when combined with sound statistical inference techniques. In light of the extensive-duration experimental design employed in our study centered on relative clause comprehension via fMRI, we endeavored to implement the GC analysis on single-trial BOLD time courses. Furthermore, we explored the correlation between the strength of GC connectivity and the complexity levels inherent to each of the relative clauses, shedding light on the intricate neural underpinnings of language comprehension.

In our endeavor, we primarily utilized a time lag of four for trial-by-trial computations, following a comprehensive assessment of the Akaike Information Criterion (AIC) and Bayesian Information Criterion (BIC), alongside rigorous physical computations of Granger causality (GC). Grounded in these methodological foundations, we delved into an exploration of single-trial GC during the comprehension of sentences featuring varying levels of complexity, with the aim of unraveling the intricate interactions among the pivotal brain regions integral to sentence processing.

The initial step involved quantifying directional influences between bivariate data extracted from the LIFG and the LSTG, precisely synchronized with the onset of each sentence. This rigorous examination yielded a wealth of GC information, encompassing the direction and strength of connectivity, effectively encapsulating the nuanced dynamics of the BOLD time course for each sentence.

To validate our observed connectivity patterns between the LIFG and the LSTG, we conducted a meticulous analysis. This entailed assessing the significance of differences between the GC outcomes—specifically, the directional connectivity from $F_{LIFG \rightarrow LSTG}$ compared to $F_{LSTG \rightarrow LIFG}$—across various sentence conditions, employing a Wilcoxon Signed Ranks test, in line with established methodologies outlined by Roebroeck et al. in 2005.

Our analytical journey culminated in the computation of correlations between sentence complexity, as ascertained through eye-tracking measures, and the strength of directional connectivity, all meticulously evaluated for each sentence. This comprehensive exploration, in turn, allowed us to gain profound insights into the intricate relationship between sentence complexity and brain connectivity. We scrutinized whether, and in what manner, the effects of sentence complexity influenced directional connectivity, unraveling the intriguing interplay between the LIFG and the LSTG, and its alignment with behavioral indices of sentence complexity across diverse sentence types.

The outcomes of our investigation unveiled compelling insights into the neural dynamics associated with sentence processing. Through the application of the Wilcoxon Signed Ranks test, we discerned that during the comprehension of SRC sentences, the strength of connectivity originating from the LIFG to the LSTG exhibited a significant predominance over the converse direction ($Z = -2.192$, $p = 0.028$). In contrast, when processing ORC sentences, no such discernible difference in connectivity strength emerged ($Z = -0.464$, $p = 0.643$).

Moreover, our investigation delved into the eye-tracking patterns of participants, unearthing a fascinating pattern. The challenges posed during the processing of Chinese SRC sentences materialized promptly upon encountering the embedded clause and endured throughout the journey until reaching the matrix subject, thereby completing the comprehension process. Thus, we turned our attention to specific regions encompassing the embedded clause (phrases 1 and 2) and the head noun (phrase 4), which consistently presented reliable differences between SRC and ORC sentences. These regions were harnessed as quantitative indicators of sentence complexity. In essence, the complexity of each sentence was quantified by aggregating the number of fixations or the cumulative fixation time allocated to the embedded clause and head noun regions. This innovative approach allowed us to gain deeper insights into the intricate nature of sentence comprehension and the factors influencing it.

Following a meticulous assessment of sentence complexity using eye movement data, we proceeded to explore the intricate relationship between sentence complexity and the strength of effective connectivity between the LIFG and the LSTG at a single-sentence level. The findings uncovered a notable revelation regarding SRC sentences. Irrespective of whether sentence complexity was quantified by the total number of fixations ($r = 0.3145$, $p = 0.0182$) or the total fixation time ($r = 0.2997$, $p = 0.0249$), a significantly positive correlation was observed between the strength of effective connectivity from the LIFG to the LSTG and sentence complexity, as illustrated in Fig. 8.6. Conversely, during the comprehension of Chinese ORC sentences, we did not observe any substantial correlation when assessing complexity through the total number of fixations ($r = 0.1659$, $p = 0.2217$) or the total fixation time ($r = 0.1444$, $p = 0.2882$).

Moreover, it is noteworthy that the effective connectivity of the opposite direction, from the LSTG to the LIFG, for both SRC and ORC sentences, did not exhibit a significant correlation with sentence complexity, regardless of whether it was determined by the total number of fixations (SRC: $r = 0.1143$, $p = 0.4016$; ORC: $r =$

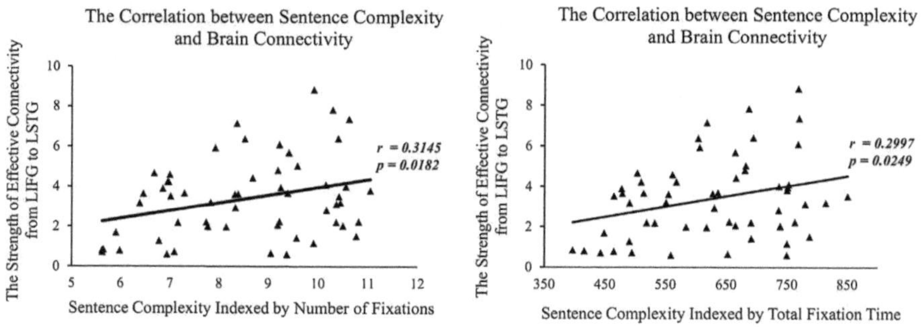

Fig. 8.6 SRC sentence complexity is positively correlated with the directional connectivity from the LIFG to the LSTG (reproduced from Xu et al., 2020b)

0.1387, $p = 0.3081$) or the total fixation time (SRC: $r = 0.0308$, $p = 0.8219$; ORC: $r = -0.1677$, $p = 0.2166$). These intricate findings deepen our understanding of the dynamic interplay between sentence comprehension and neural connectivity in distinct linguistic contexts.

8.9 Conclusion

In the pursuit of illuminating the nature of sentence comprehension in the context of Chinese language processing, we embark on a comprehensive overview of a series of neuroimaging experiments. Our objective is to substantiate the observed ORC advantage, where SRC sentences prove to be relatively more challenging to comprehend in Chinese. This conclusion is grounded in a robust body of convergent evidence drawn from multimodal functional neuroimaging results.

A pivotal facet of this evidence hinges on eye movement analysis, which aligns seamlessly with previous research findings, including noteworthy studies such as those conducted by Gibson and Wu (2013), Hsiao and Gibson (2003), Qiao et al. (2012), Sung et al. (2016), Xu et al. (2019b), and Xu et al. (2020a). These collective findings consistently demonstrate that Chinese SRC sentences present a heightened level of complexity in processing, as evidenced by the emergence of larger numbers of fixations or extended durations of fixations, particularly within the embedded clause and head noun regions. Significantly, Chinese, characterized by its head-final language structure where the head noun trails the embedded clause, forms the foundation for the unique challenges presented by SRC sentences. Their non-canonical word order necessitates increased integration demands, leading to prolonged reading times. This is largely due to the requirement to sustain incomplete dependencies across an extended filler-gap distance until the mapping of thematic roles to arguments is accomplished. This, in contrast, stands in stark contrast to the relative ease associated with reading ORC sentences featuring a canonical word order.

It is noteworthy that, within our works, the influence of sentence complexity on the total fixation number, particularly within the embedded clause and head noun regions, was notably robust, with a more pronounced impact compared to total fixation time. Nonetheless, the overarching pattern of total fixation time also aligns harmoniously with theoretical expectations. As such, the quantifiable disparities in complexity across individual sentences are meticulously reflected in the sum of real-time eye-tracking indices, carefully observed in specific regions of the sentences. These eye-tracking indices serve as a valuable behavioral reference point for the exploration of corresponding neural mechanisms that underpin the intricate process of sentence comprehension.

The identical set of relative clause materials employed in our eye-tracking experiment served as the foundation for a separate fMRI investigation involving a distinct cohort of healthy normal volunteers. In the initial analysis of the fMRI dataset, we adopted a conventional GLM data analysis framework, specifically SPM8. Given the inherently comparable processes underlying the comprehension of Chinese subject-extracted relative clause (SRC) and object-extracted relative clause (ORC) sentences, the active functional brain areas associated with these processes exhibited remarkable overlap. Consequently, the outcomes of the contrast analysis revealed a limited number of surviving brain areas, as previously documented (Xu et al., 2020a).

Furthermore, in the context of directional Granger causality analysis, limited causal influence was identified, specifically from the LIFG to the LSTG. However, our adoption of an fMRI data analysis framework rooted in Independent Component Analysis (ICA) yielded robust and consistent activations within the LIFG and the LSTG across all participants. This discovery aligns harmoniously with earlier neuroimaging studies that have underscored the pivotal role of the LIFG and the LSTG in the comprehension of relative clause (RC) sentences. Previous studies, such as Bornkessel-Schlesewsky and Schlesewsky (2013), Constable et al. (2004), Friederici (2003, 2011), Hagoort (2014), Just et al. (1996), Kuhnke et al. (2017), Meyer et al. (2012), Pattamadilok et al. (2016), and our previous work (Xu et al., 2020a, 2020b), have consistently highlighted the significance of these regions.

This confluence of findings underscores the reliability and consistency of the language comprehension circuit, a robust neuroimaging signature that transcends various functional neuroimaging studies, and reaffirms the essential roles played by the LIFG and the LSTG in the intricate process of RC sentence comprehension.

Subsequent to our initial examination, Granger causality analysis was systematically applied to the outcomes of the traditional GLM-based fMRI data analysis. This focused exploration unveiled a discernible but restricted unidirectional connectivity pattern, specifically observed from the LIFG to the LSTG, exclusively in the context of SRC sentence comprehension, as previously reported (Xu et al., 2020a). It is imperative to recognize, however, that this analysis was subject to a fundamental limitation—the pronounced interdependence of the preprocessed BOLD time courses within these two vital brain regions, as elucidated in Sect. 8.4. To enable the Granger causality (GC) analysis, a prerequisite was to disentangle and account for the inherent correlations present within the BOLD time courses originating from these two critical regions. It is conceivable that this inherent correlation presented a significant challenge, serving as the primary factor underpinning the constrained Granger causality findings.

In pursuit of a more refined and sensitive approach to Granger causality (GC) analysis, with the added ambition of achieving connectivity resolution at the level of individual trials, we embarked on an Independent Component Analysis (ICA) decomposition of the fMRI BOLD data. This meticulous decomposition process served the purpose of segregating the data into distinct and independent BOLD components. Subsequently, we engaged in a component selection procedure grounded in spatial distribution, culminating in the identification and selection of two key components: the LIFG component and the LSTG component. This component selection process was meticulously executed for each individual subject's ICA results. This strategic course of action was undertaken with the overarching goal of preserving the requisite degrees of freedom for conducting GC analysis on the BOLD time courses, which emanated from the carefully selected ICA components. This approach offers a potential solution to the inherent challenges of GLM model-based brain activation analysis, particularly those stemming from the utilization of smoothed time courses to identify brain BOLD activations. Moreover, this method facilitated the unprecedented prospect of performing GC analysis on the epoched BOLD time courses, dissecting the neural dynamics at a granular level, trial by trial, across various sentence comprehension tasks.

The results of our study shed light on the distinct neural dynamics underlying the comprehension of Chinese ORC sentences, which are relatively straightforward to understand, and SRC sentences, characterized by heightened integration demands. In the case of ORC sentences, characterized by their canonical word order, comprehension primarily necessitates a unidirectional connection from the LIFG to the LSTG. This connectivity primarily supports the integration of lexical-syntactic information and serves as a sufficient mechanism for sentential integration, given the shorter distances involved in memory retrieval for sequentially integrated word meanings.

Conversely, the comprehension of SRC sentences, marked by their non-canonical word order and the resultant greater integration complexity, entails a more intricate connectivity pattern. In addition to the connection from the LIFG to the LSTG, SRC sentence comprehension requires connectivity from the LSTG to the LIFG as well. This bidirectional interaction is indicative of the intensified integration demands for reconstructing argument order, particularly in the context of comprehending "who did what to whom."

Based on these observed patterns of effective connectivity in SRC and ORC sentence comprehension, it is reasonable to infer that the integration of lexical-syntactic information, a fundamental process in sentence comprehension, is commonly facilitated by the neural network extending from the LIFG to the LSTG. This connectivity emerges as a pivotal element in the comprehension of relative clause sentences as a whole. As a result, it should be more reasonable to see this directional connectivity even in comprehending the ORC sentences as compared to the previously mentioned GC analysis results based on the GLM model-based activation analysis.

Consistent with prior studies on visual and verbal memory, the sustained activation of representations in the LSTG is subject to top-down modulation by frontal regions (Fiebach et al., 2001; Sreenivasan et al., 2014). Likewise, previous research in semantic priming and speech perception tasks has highlighted the causal top-down influence of the frontal cortex on the temporal cortex (Cope et al., 2017; Hartwigsen et al., 2017). Our findings are in harmony with these earlier observations, emphasizing the role of the LIFG in exerting top-down control on the LSTG, thereby facilitating the integration of upcoming words critical for sentence comprehension.

Additionally, our investigation unveiled a noteworthy observation: the intricacy of sentence structures could be more effectively elucidated by considering the Granger causality influence from the LIFG to the LSTG than the reverse direction. As illustrated in Fig. 8.6, there was a discernible relationship between sentence complexity and integration demands in the comprehension of SRC sentences. Consequently, more complex SRC sentences demanded a stronger effective connectivity from the LIFG to the LSTG for successful comprehension. These outcomes convey several critical insights.

Firstly, they suggest that the strength of connectivity from the LIFG to the LSTG closely mirrors the processing intricacies associated with integrating lexical-syntactic information, a phenomenon substantiated by behavioral measures such as prolonged and more frequent eye fixations.

Secondly, the presence of a significant correlation exclusively in the context of SRC sentences, and not in ORC sentences, implies that the integration of lexical-syntactic information becomes markedly more demanding when compounded with syntactic processing challenges, a hallmark of SRC sentences. In essence, the integration of word meanings within a noncanonical word order necessitates substantial cognitive effort due to the need to reconstruct the argument hierarchy and grasp the relationships among "who is doing what to whom." Consequently, the effective connectivity from the LIFG to the LSTG, attributed to the integration process, dynamically adapts to the heightened integration demands.

It is imperative to acknowledge that our analysis does not purport to present a comprehensive model of visual sentence comprehension; instead, it isolates key nodes crucial to the process of sentential integration. Therefore, it is plausible that these areas engage in interactions with additional brain regions not encompassed in our analysis, thus warranting further research to explore this possibility.

Moreover, it prompts intriguing inquiries into individual differences in sentence comprehension. Future studies might investigate whether the classification of LIFG and LSTG activation into the same or distinct components correlates with individual variations in reading ability. We anticipate that our proposed data analysis framework could offer a viable avenue for investigating these hypotheses and expanding our comprehension of the neural underpinnings of sentence processing.

References

Baayen, R. H., Davidson, D. J., & Bates, D. M. (2008). Mixed-effects modeling with crossed random effects for subjects and items. *Journal of Memory and Language, 59*(4), 390–412. https://doi.org/10.1016/j.jml.2007.12.005

Bornkessel-Schlesewsky, I., & Schlesewsky, M. (2013). Reconciling time, space and function: A new dorsal-ventral stream model of sentece comprehension. *Brain and Language, 125*(1), 60–76. https://doi.org/10.1016/j.bandl.2013.01.010

Brovelli, A. (2012). Statistical analysis of single-trial Granger causality spectra. *Comp. Mathemat. Method Med., 1–10*,. https://doi.org/10.1155/2012/697610

Constable, R. T., Pugh, K. R., Berroya, E., Mencl, W. E., Westerveld, M., Ni, W., et al. (2004). Sentence complexity and input modality effects in sentence comprehension: An fMRI study. *NeuroImage, 22*(1), 11–21. https://doi.org/10.1016/j.neuroimage.2004.01.001

Cope, T. E., Sohoglu, E., Sedley, W., Patterson, K., Jone, P. S., Wiggins, J., et al. (2017). Evidence for causal top-down frontal contributions to predictive processes in speech perception. *Nature Communications, 8*(1), 2154. https://doi.org/10.1038/s41467-017-01958-7

Deshpande, G., LaConte, S., James, G. A., & Hu, S. P. (2009). Multivariate granger causality analysis of fMRI data. *Hum. Brain Mapp, 30*(4), 1361–1373. https://doi.org/ https://doi.org/10.1002/hbm.20606

Duann, J.-R., Ide, J. S., Luo, X., & Li, C.-S.R. (2009). Functional connectivity delineates distinct roles of the inferior frontal cortex and presupplementary motor area in stop signal inhibition. *Journal of Neuroscience, 29*(32), 10171–10179. https://doi.org/10.1523/JNEUROSCI.1300-09.2009

Fiebach, C. J., Schlesewsky, M., & Friederici, A. D. (2001). Syntactic working memory and the establishment of filler-gap dependencies: Insights from ERPs and fMRI. *Psychological Research, 30*(3), 321–338. https://doi.org/10.1023/A:1010447102554

Friederici, A. D., Rüschemeyer, S.-A., Hahne, A., & Fiebach, C. J. (2003). The role of left inferior frontal and superior temporal cortex in sentence comprehension: Localizing syntactic and semantic processes. *Cerebr. Cortex, 13*(2), 170–177. https://doi.org/10.1093/cercor/13.2.170

Friederici, A. D. (2011). The brain basis of language processing: From structure to function. *Physiological Reviews, 91*(4), 1357–1392. https://doi.org/10.1152/physrev.00006.2011

Hagoort, P. (2014). Nodes and networks in the neural architecture for language: Broca's region and beyond. *Current Opinion in Neurobiology, 28*, 136–141. https://doi.org/10.1016/j.conb.2014.07.013

Hartwigsen, G., Henseler, I., Stockert, A., Wawrzyniak, M., Wendt, C., Klingbeil, J., et al. (2017). Integration demands modulate effective connectivity in a fronto-temoporal network for contextual sentence integration. *NeuroImage, 147*, 812–824. https://doi.org/10.1016/j.neuroimage.2016.08.026

Gibson, E., & Wu, H. H. I. (2013). Processing Chinese relative clauses in context. *Language & Cognitive Processes, 28*(1–2), 1–31. https://doi.org/10.1080/01690965.2010.536656

Goebela, R., Roebroecka, A., Kimb, D.-S., & Formisano, E. (2003). Investigating directed cortical interactions in time-resolved fMRI data using vector autoregressive modeling and Granger causality mapping. *Magn. Reson. Imag., 21*(10), 1251–1261. https://doi.org/10.1016/j.mri.2003.08.026

Granger, C. W. J. (1969). Investigating causal relations by econometric models and crossspectral methods. *Econometrica, 37*(3), 424–438. https://doi.org/10.2307/1912791

Hsiao, F., & Gibson, E. (2003). Processing relative clauses in Chinese. *Cognition, 90*(1), 3–27. https://doi.org/10.1016/s0010-0277(03)00124-0

Jager, L., Chen, Z., Li, Q., Lin, C.-J.C., & Vasishth, S. (2015). The subject-relative advantage in Chinese: Evidence for expectation-based processing. *Journal of Memory and Language, 79–80*, 97–120. https://doi.org/10.1016/j.jml.2014.10.005

Just, M. A., Carpenter, P. A., Keller, T. A., Eddy, W. F., & Thulborn, K. R. (1996). Brain activation modulated by sentence comprehension. *Science, 274*(5284), 114–116. https://doi.org/10.1126/science.274.5284.114

King, J., & Just, M. A. (1991). Individual differences in syntactic processing: The role of working memory. *Journal of Memory and Language, 30*(5), 580–602. https://doi.org/10.1016/0749-596X(91)90027-H

Kuhnke, P., Meyer, L., Friederici, A. D., & Hartwigsen, G. (2017). Left posterior inferior frontal gyrus is causally involved in reordering during sentence processing. *NeuroImage, 148*, 254–263. https://doi.org/10.1016/j.neuroimage.2017.01.013

Meyer, L., Obleser, J., Anwander, A., & Friederici, A. D. (2012). Linking ordering in Broca's area to storage in left temporo-parietal regions: The case of sentence processing. *NeuroImage, 62*(3), 1987–1998. https://doi.org/10.1016/j.neuroimage.2012.05.052

Pattamadilok, C., Dehaene, S., & Pallier, C. (2016). A role for left inferior frontal and posterior superior temporal cortex in extracting a syntactic tree from a sentence. *Cortex, 75*, 44–55. https://doi.org/10.1016/j.cortex.2015.11.012

Qiao, X., Li., Shen, & Forster, K. (2012). Relative clause processing in Mandarin: Evidence from the maze task. *Language & Cognitive Processes, 27*(4), 611–630. https://doi.org/10.1080/01690965.2011.578394

R Core Team (2011). A language and environment for statistical computing. 3–900051–07–0. R Foundation for Statistical Computing, Vienna, Austria.

Roebroeck, A., Formisano, E., & Goebel, R. (2005). Mapping directed influence over the brain using Granger causality and fMRI. *NeuroImage, 25*(1), 230–242. https://doi.org/10.1016/j.neuroimage.2004.11.017

Sato, J. R., Takahashi, D. Y., Arcuri, S. M., Sameshima, K., Morettin, P. A., & Baccala, L. A. (2009). Frequency domain connectivity identification: An application of partial directed coherence in fMRI. *Human Brain Mapping, 30*(2), 452–461. https://doi.org/10.1002/hbm.20513

Sreenivasan, K. K., Curtis, C. E., & D' Esposito, M. (2014). Revisiting the role of persistent neural activity during working memory. *Trends Cognit. Sci., 18*(2), 82–89. https://doi.org/10.1016/j.tics.2013.12.001

Sung, Y.-T., Cha, J.-H., Tu, J.-Y., Wu, M.-D., & Lin, W.-C. (2016). Investigating the processing of relative clauses in Mandarin Chinese: Evidence from eye-movement data. *Journal of Psycholinguistic Research, 45*(5), 1089–1113. https://doi.org/10.1007/s10936-015-9394-y

Xu, K., Duann, J. R., Hung, D. L., & Wu, D. H. (2019a). Preference for object relative clauses in Chinese sentence comprehension: Evidence from online self-paced reading time. *Frontiers in Psychology, 10*, 2210. https://doi.org/10.3389/fpsyg.2019.02210

Xu, K., Duann, J. R., & Wu, D. H. (2019b). Preference for object relative clauses in Chinese sentence comprehension: Evidence from online self-paced reading time. *Frontiers in Psychology, 20*, 1–15.

Xu, K., Wu, D. H., & Duann, J. R. (2020a). Enhanced left inferior frontal to left superior temporal effective connectivity for complex sentence comprehension: FMRI evidence from Chinese relative clause processing. *Brain and Language, 200*, 1–10.

Xu, K., Wu, D. H., & Duann, J. R. (2020b). Dynamic brain connectivity attuned to the complexity of relative clause sentences revealed by a single-trial analysis. *NeuroImage, 217*, Article 116920. https://doi.org/10.1016/j.neuroimage.2020.116920

Chapter 9
Neural Effects of Age-of-Acquisition and Word Frequency in Object Naming

Wen-Jui Kuo, Hsin-Ju Lee, and Fa-Hsuan Lin

Abstract In this study, we examined the neural effects of the age-of-acquisition (AoA) of an object's meaning and the word frequency of the object's name during picture naming using fMRI. Our findings identified overlapping brain areas associated with these two effects, including the inferior frontal cortex, the supplementary motor area (SMA), and the left occipitotemporal cortex. This overlap suggests that the processes underpinning these effects are regulated by both AoA and word frequency. Spatially distinct areas were also identified. The involvement of the middle temporal gyrus in the AoA effect underscores its critical role in lexical-semantic processing or retrieval. The engagement of subcortical structures and the left inferior parietal cortex in the word frequency effect suggests its influence on the construction of lexical phonology. Furthermore, the study revealed different neural effects of word frequency on naming objects with early and late AoA. This differentiation may shed light on how the organization of conceptual content interacts with reading processes that are sensitive to word frequency.

Keywords Age-of-acquisition (AoA) · Word frequency · Object naming · fMRI

9.1 Introduction

Object naming is a complex psychological task that involves processes such as object recognition, semantic and lexical retrieval, and phonological encoding for articulation (Levelt, 1999). This task, as a form of speech production, has frequently been used to explore knowledge representation and organization in the brain. It's clear

W.-J. Kuo (✉)
Institute of Neuroscience, National Yang Ming Chiao Tung University, Taipei, Taiwan
e-mail: wjkuo@nycu.edu.tw

Brain Research Center, National Yang Ming Chiao Tung University, Taipei, Taiwan

H.-J. Lee · F.-H. Lin
Physical Sciences Platform, Sunnybrook Research Institute, Toronto, ON, Canada

Department of Medical Biophysics, University of Toronto, Toronto, ON, Canada

D. H. Wu (ed.), *Cognitive and Neural Foundations of Chinese Reading*, Chinese Language Learning Sciences, https://doi.org/10.1007/978-981-96-6669-0_9

that throughout our developmental journey, we acquire concepts of different objects at different ages of our lives. Evidence indicates that the age at which we learn about an object—its age-of-acquisition (AoA)—can influence the efficiency of these cognitive processes. Specifically, objects learned early in life are recognized and articulated more swiftly than those learned later (Carroll & White, 1973). The AoA effect is a well-documented phenomenon across numerous studies, leading to various hypotheses regarding its mechanisms, such as impacts on object recognition, the phonological output lexicon, and the lexical semantic system (e.g., Brown & Watson, 1987; Brysbaert, 1996; Ellis & Lambon Ralph, 2000; Gerhand & Barry, 1999). In this study, we delve into the neural dimensions of the AoA effect through functional magnetic resonance imaging (fMRI), aiming to illuminate the neural pathways through which our brains organize and retrieve semantic concepts.

To investigate the AoA effect, it is crucial to recognize that concepts acquired earlier often have names with higher word frequency than those acquired later. This observation has led to debates on whether AoA effects are similar to those of word frequency (Balota et al., 2004; Zevin & Seidenberg, 2004). Researchers question whether these two factors have a unified impact on lexical organization. Beyond the phonological explanation of the AoA effect proposed by earlier studies (Brown & Watson, 1987; Morrison & Ellis, 1995), recent analyses suggest that the AoA effect may be either related to frequency or occur independently (Belke et al., 2005; Brysbaert & Ghyselinck, 2006). In tasks like word naming and lexical decision-making, the magnitudes of AoA and word frequency effects are strongly correlated. Conversely, in object naming tasks, which require the selection or mapping of specific concepts, the AoA effect is significantly more pronounced. Thus, this study not only examines the frequency-associated AoA effect but also explores a potential frequency-independent AoA effect, aiming to elucidate the neural correlates of AoA and word frequency in object naming.

The neural impact of AoA has been extensively explored through a variety of neuroimaging studies utilizing diverse tasks and materials, such as word reading and object naming (Ellis et al., 2006; Fiebach et al., 2003; Hernandez & Fiebach, 2006; Hernandez & Li, 2007). For instance, studies have shown that early acquired words elicit higher activity in the precuneus for both written and spoken forms, whereas late learned words are associated with increased activity in the left inferior frontal cortex and anterior insula. Notably, the left inferior frontal activity related to AoA is spatially distinct from the areas linked to the word frequency effect, which are found more dorsally in the left inferior frontal cortex (Fiebach et al., 2002, 2003). Further, an fMRI study by Hernandez and Fiebach (2006) demonstrated that late learned words prompted higher activity in the left temporal cortex, right inferior frontal cortex, and certain subcortical structures when visually presented for reading. In a study conducted by Ellis et al. (2006) on object naming, early-learned concepts were associated with higher activity in the occipital and temporal regions, while late-learned items induced higher activity in the left fusiform gyrus. This suggests that early-learned concepts foster a more detailed visual and rich semantic representation, necessitating a greater effort in mapping between visual and semantic representations for late-learned items. However, electrophysiological evidence, such as the findings from Perret et al. (2014), indicates that phonological word-form encoding is the critical stage at which the AoA effect manifests during object naming. This overview highlights that while significant research has been conducted on the neural effects of AoA, findings vary, pointing to the complex nature of how AoA influences brain activity.

Imaging studies have been conducted to explore the neural effects of word frequency during object naming tasks (Grande et al., 2011; Graves et al., 2007). The study by Graves et al. (2007) identified neural correlates of the word frequency effect in object naming, which included the posterior superior temporal gyrus, posterior inferior temporal gyrus, and the inferior frontal gyrus within the left hemisphere. Similarly, Grande et al. (2011), in a study aimed at children, also reported a word frequency effect in the left inferior frontal gyrus. These findings suggest that these areas play a crucial role in retrieving the lexical phonology corresponding to the depicted semantic concept (Jescheniak & Levelt, 1994), with both studies pinpointing the left inferior frontal gyrus as a common site associated with the word frequency effect in object naming. Electrophysiological studies further support this, indicating that the word frequency effect, which is linked to lexical access and the retrieval of lexical word forms from the mental lexicon, occurs approximately 200 ms after stimulus onse. Collectively, these studies underscore that the neural foundation of lexical access during spoken word production, influenced by word frequency, is predominantly facilitated by mechanisms located in the inferior frontal gyrus. This aligns with previous research on the neural basis of the word frequency effect observed in reading Chinese and English, highlighting a consistent neural pattern across different languages and tasks.

This study is designed to explore the neural impacts of AoA and word frequency during object naming tasks. The simultaneous examination of these effects within the same research framework has not been previously conducted, making the outcomes of this study crucial for elucidating how the brain organizes concepts developed over time and addresses the effects arising from the word frequency with which a concept accumulates. To achieve this goal, we will analyze brain activity associated with the word frequency effect under different AoA conditions, specifically early and late AoA. Connectionist models suggest that patterns of neural connections for items learned later are challenging to structure effectively due to decreased neural network plasticity, a consequence of prior learning of other items (Ellis & Lambon Ralph, 2000). We therefore expect that neural effects of word frequency of early and late AOA would be different and hosted by differentiable neural networks. Additionally, existing research highlights that AoA can exert its effect at the stage of object recognition (Catling & Johnston, 2009; Catling et al., 2008; Dent et al., 2007). If the AoA effect is observed in the occipital or ventral visual processing streams, it would substantiate the connection between this neural evidence and behavioral observations at a neural level.

9.2 Methods

9.2.1 Participants

There are twenty participants (13 females; mean age: 23.4 years, ranging from 22 to 26) recruited for the experiment. All participants are right-handed, native-Chinese speakers with no history of neurological disorders, and had normal or corrected-to-normal vision during the experiment. Handedness of the participants was verified using the Edinburgh Inventory (Oldfield, 1971). Informed consent of all participants was obtained before MR scanning, with the protocol approved by the Institutional Review Board of National Yang Ming Chiao Tung University.

9.2.2 Material and Experimental Task

The experimental stimuli are black-and-white 2-D line drawings of objects that are drawn from a larger corpus of picture stimuli (the Center for Research in Language, International Object naming Project, CRL-IPNP) (Szekely et al., 2003). The objects all have a name which is a 2-character compound word, i.e., two syllables for their word length, by which the word length is well under control for naming. AoA of the picture stimuli was previously collected with a group of 128 college students tasked with behavioral rating to the corpus. Word frequency of the object's name is obtained from a Chinese word corpus. Based on the rating and corpus data, two AoA levels (early and late) and two word frequency levels (high and low) are adopted (see Table 9.1 for the details). For familiarization, every pictured object and its name were paired and presented to the participants before MRI scanning.

During MRI scanning, the participants were fixed to do a picture naming task. Every naming trial includes a fixation presentation for 500 msc and, then, a picture presentation for 500 msc for naming. The participants were asked to name the pictured objects as soon and as accurate as possible. Because the MR system was not equipped to record naming responses, we could not fully ensure that the participants were engaged in the naming task all the time. Therefore, among the naming trials, we

Table 9.1 Characteristics of the stimuli

	Early AoA		*Late AoA*	
	Low WF	*High WF*	*Low WF*	*High WF*
AoA (years)	5.51 (0.629)	5.24 (0.704)	8.18 (1.148)	7.31 (0.891)
WF (per million)	2.95 (1.722)	25.22 (22.573)	1.45 (0.918)	16.6 (8.023)
Familiarity rating	5.53 (0.687)	5.85 (0.589)	4.41 (0.79)	1.21 (0.189)

Note Numbers indicate the mean and standard deviation (in parentheses) of the parameters. AoA, age-of-acquisition. WF, word frequency. Familiarity rating, from low to high, ranges from 1 to 7

arranged some simple target (a white circle) detection trials where the participants needed to press a button to notice it. If detection performance is poor, (i.e., $< 85\%$), we will assume that the participant will miss a certain proportion of the naming trials as well. The fMRI data of the participant will be discarded from the analysis. Altogether, there were 108 picture naming trials, 126 null trials for jittering inter-trial interval, and 72 simple detection trials for monitoring the participants' engagement. All trials were randomly ordered.

9.3 MRI Procedure

MR scanning was performed that used a 3 T MRI (Tim Trio, Siemens, Erlangen, Germany) interfaced with a 32-channel phased-array head coil. A T2*-weighted gradient-echo echo planar imaging (EPI) sequence was used for fMRI scanning, with slice thickness = 3.4 mm, in-plane resolution (64×64) = 3.44×3.44 mm, and TR/TE/θ = 2000 ms/30 ms/90°. Thirty-three axial slices were acquired to cover the whole brain. In addition to the first three volumes for signal equilibrium, for each slice, 306 images were acquired. The anatomical, T1-weighted high-resolution image ($1 \times 1 \times 1$ mm) was acquired using a standard MPRAGE sequence (TR/TE/TI = 2530/3.49/1100 ms, flip angle = 7°).

9.3.1 Data Processing and Statistical Analysis

There were two participants' data sets excluded from the analysis due to poor discrimination performance. Eighteen fMRI data sets were analyzed using SPM8 (Welcome Department of Imaging Neuroscience). MR images were temporally corrected for slice timing, spatially corrected for motion, normalized to MNI space, and spatially smoothed with an 8 mm FWHM Gaussian kernel. The resulting time-series was high-pass-filtered with a cut-off time window of 128 s. A two-level approach for statistical analysis was adopted. For each participant, there were five event types to be modelled, including four event types from the 2 (AoA) $\times$ 2 (word frequency) arrangement and one event type for the detection trials. Six motion parameters were included as regressors. An image for each trial type was computed by fitting each voxel time-series with the known time-series of the five event types convolved with a canonical hemodynamic response function with a time derivative. Unless stated otherwise, all effects of interest were reported at voxel $p < 0.001$ and corrected for multiple comparisons at the cluster level using *FDR* $p < 0.05$.

9.4 Results

9.4.1 Behavioral Results

In the fMRI experiment, only the simple target detection responses were recorded. The mean detection rate of the simple target detection task was 0.98, and the average reaction time was 546 ms. The accuracy of two participants was lower than 85%, and their image data were excluded from further analysis.

9.4.2 Functional MRI Results

Picture naming activated a number of brain areas, including the bilateral occipital, parietal, temporal, and frontal cortices, which was consistent with the previous fMRI studies on language processing (Ellis et al., 2006; Gronholm et al., 2005; Martin et al., 1996; Price, 2000, 2012; van Turennout et al., 2000, 2003). Activation was also observed in subcortical brain areas, such as the thalamus and basal ganglia. Neural effects specific to the AoA and word frequency in object naming were presented separately in the following.

9.4.3 Neural Effects of the AoA

Brain areas associated with the AoA effect, i.e., higher activity for the late-AoA than for the early-AoA items, were bilaterally distributed, including the pre-motor gyrus, inferior frontal gyrus, SMA, middle temporal gyrus, and visual cortices extending from primary to associative areas (see Fig. 9.1A). Several areas in the frontal cortex overlapped with those related to the word frequency effect (see Fig. 9.2 and Table 9.2).

9.4.4 Neural Effects of the Word Frequency

Brain areas associated with the word frequency effect, i.e., activity higher for low word frequency than for high word frequency, included left pre-motor and inferior frontal cortex, SMA, superior parietal cortex, and visual association cortex (see Fig. 9.1B). Subcortical structures, such as the caudate and thalamus, were included as well. Several areas in the frontal lobe overlapped with those related to the AoA effect (see Fig. 9.2 and Table 9.3).

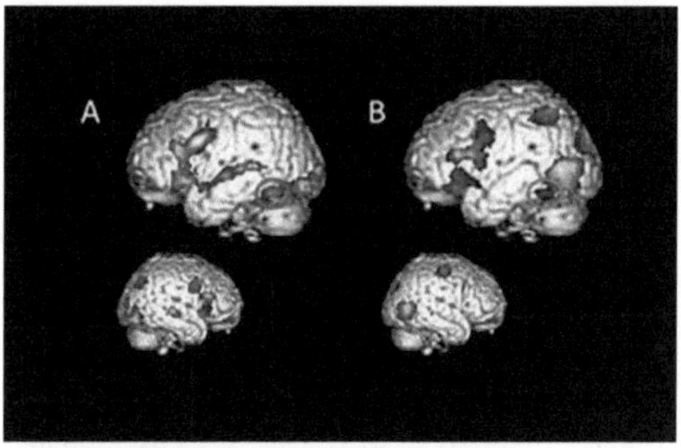

Fig. 9.1 The statistical parametric maps of the AoA effect (A) and the word frequency effect (B). The noted areas survive the threshold $p = 0.001$, corrected at the cluster level using FDR $p = 0.05$

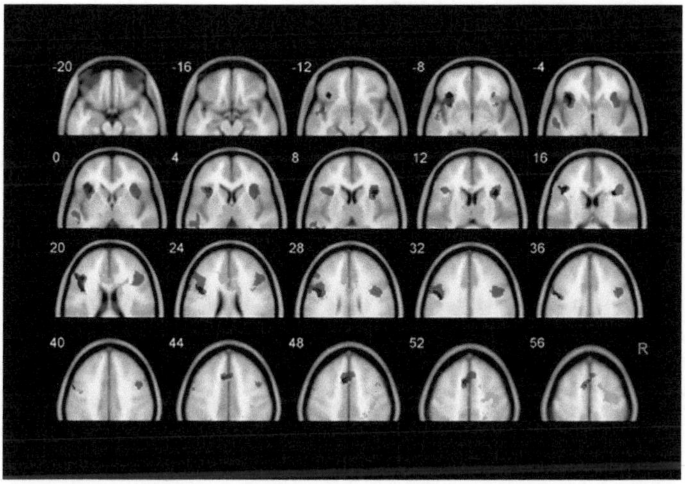

Fig. 9.2 Brain areas associated with the AoA and word frequency effects were colored in red and green to denote the overlaps. The noted areas survive the threshold $p = 0.001$, corrected at the cluster level using FDR $p = 0.05$

9.4.5 Neural Effects of the Word Frequency for the Early-AoA Items

For the early-AoA category, naming objects with lower word frequency resulted in higher activities, more specifically, in the ventral part of inferior frontal gyrus and left associative visual cortex (see Fig. 9.3A and 3C in yellow and Table 9.4).

Table 9.2 Peaks of the networks revealed by contrasting images of late-AoA vs. early AoA

	Left hemisphere				Right hemisphere			
Brain areas	x	y	z	*Z-value*	x	y	z	*Z-value*
Precentral G	−46	−2	32	**3.82**	44	0	34	**4.14**
Inferior frontal G	−34	26	−6	**4.94**	32	28	−6	**3.88**
	−60	12	28	**4.34**				
	−44	8	28	**4.02**				
Middle frontal G	−50	30	24	**3.95**	46	20	20	**4.2**
Insula	−32	20	4	**3.97**	32	22	6	**5.07**
Superior/ Medial frontal G	−8	20	46	**3.91**	16	8	62	**3.79**
	−8	14	48	**3.83**	2	22	52	**3.75**
Superior temporal G	−66	−40	6	**3.69**	50	−22	−4	**3.79**
	−52	−30	4	**3.69**				
	−54	−16	−2	**3.92**				
	−46	6	−12	**3.44**				
	−52	0	−10	**3.43**				
Middle temporal G	−48	−40	6	**3.59**	64	−26	2	**3.98**
	−58	−6	−4	**3.28**				
Superior pariteal L	−26	−60	40	**3.69**				
Precuneus	−24	−70	38	**3.67**	34	−68	38	**3.83**
Fusiform G	−46	−60	−18	**4.06**				
Inferior occipital G	−14	−92	−8	**5.07**	22	−88	−10	**4.79**
Middle occipital G	−22	−86	−8	**4.73**	40	−74	−10	**4.48**

Note Coordinates given are MNI coordinates. Voxel-wise threshold $p < 0.001$; threshold for cluster extent FDR $p < 0.05$

9.4.6 Neural Effects of the Word Frequency for the Late-AoA Items

For the late-AoA category, naming objects with low word frequency resulted in higher activities in the dorsal part of inferior frontal gyrus, left inferior parietal

Table 9.3 Peaks of the networks revealed by contrasting images of low-frequency vs. high-frequency

	Left hemisphere				Right hemisphere			
Brain areas	x	y	z	*Z-value*	x	y	z	*Z-value*
Precentral G	−48	0	32	**4.08**	34	−14	58	**4.24**
Middle frontal G	−52	4	42	**3.59**				
	−42	24	18	**3.59**				
Inferior frontal G	−44	6	24	**3.81**				
	−46	12	22	**3.68**				
	−54	26	16	**3.93**				
	−58	14	16	**3.78**				
	−26	28	−12	**4.79**				
	−36	24	−2	**4.31**				
Insula	−40	8	14	**4.22**				
Superior/ Medial frontal G	−4	12	48	**3.94**				
Superior parietal L	−34	−54	48	**4.78**				
	−26	−52	52	**4.31**				
Middle occipital G	−50	−62	2	**5.07**	50	−72	4	**5.21**
	−46	−78	8	**5.04**	52	−64	6	**5.01**
	−42	−70	4	**4.72**	48	−66	2	**4.85**
Inferior occipital G	−40	−74	−2	**4.15**				
Fusiform G	−46	−72	−10	**3.9**				
Thalamus	−6	−6	8	**4.18**	8	−6	8	**3.95**
Caudate					16	4	24	**3.97**

Note Coordinates given are MNI coordinates. Voxel-wise threshold $p < 0.001$; threshold for cluster extent FDR $p < 0.05$

(supra-marginal) cortex, and bilateral middle occipital cortex (see Fig. 9.3B and 9.3C in blue and Table 9.4).

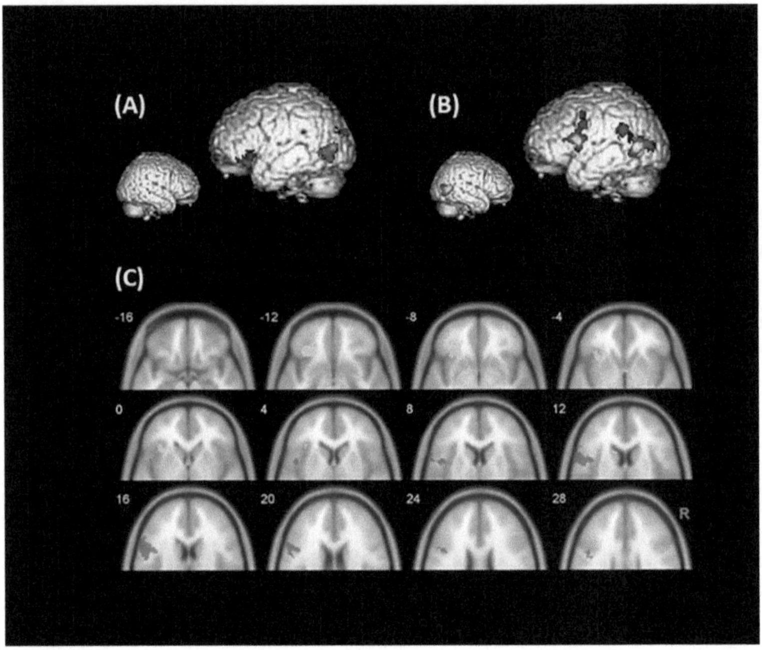

Fig. 9.3 Distinct areas associated with word frequency effect of the early AoA objects (A and yellow parts in C) and the word frequency effect of the late-AoA objects (B and blue parts in C) in the inferior frontal cortex. The noted areas survive the threshold $p = 0.001$, corrected at the cluster level using FDR $p = 0.05$

9.5 Discussion

This fMRI study aimed to investigate the neural impacts of Age of Acquisition (AoA) and word frequency during a picture naming task. By analyzing the cortical distribution of these two effects and their interaction, we discussed the underlying psychological processes.

9.5.1 The AoA Effect in the Brain

Beyond the frequently mentioned origin of the AoA effect in the phonological output lexicon, AoA also plays a role in lexical semantic processing, as evidenced by previous research (Brysbaert & Ghyselinck, 2006; Brysbaert et al., 2000; Gerhand & Barry, 1999; Morrison & Ellis, 1995; Morrison et al., 1992). This study identified several brain regions showing the AoA effect, where increased activity was observed for items learned later in life, aligning with these prior behavioral findings. Initially,

Table 9.4 Word-frequency effects for the early AoA and late AoA objects

	Left hemisphere				Right hemisphere			
Brain areas	x	y	z	*Z-value*	x	y	z	*Z-value*
WFE @ early_AoA								
Inferior frontal G	-26	28	-12	**4.54**				
	-32	32	-8	**4.29**				
Insula	-34	22	0	**3.83**				
Middle occipital G	-40	-76	6	**4.29**				
	-42	-68	2	**3.76**				
	-42	-74	-4	**3.73**				
WFE @ late_AoA								
Precentral G	-48	0	32	**3.83**				
Inferior frontal G	-60	14	16	**4.24**				
Insula	-44	8	12	**3.87**				
Supramarginal G	-60	-48	26	**3.82**				
Superior temporal G	-64	-56	20	**3.45**				
Middle occipital G	-52	-62	2	**4.45**	52	-64	6	**4.96**
	-42	-64	4	**4.06**	44	-62	4	**4.31**
	-52	-76	8	**3.88**	50	-74	8	**4.04**

Note Coordinates given are MNI coordinates. Voxel-wise threshold $p < 0.001$; threshold for cluster extent FDR $p < 0.05$

the involvement of the pre-motor, Supplementary Motor Area (SMA), and posterior inferior frontal cortex, known for their collaborative role in encoding lexical phonology (Bookheimer, 2002; Fiez & Petersen, 1998; Kuo et al., 2004; Poldrack et al., 1999; Price, 2012), indicates that processing late-learned items may require greater effort in encoding their phonological plans for articulation. Furthermore, enhanced activations in the middle temporal gyri and the ventral part of the inferior frontal gyrus suggest that late-learned concepts necessitate more intensive processing for semantic lexical retrieval and lexical control (Bookheimer, 2002; Fiez & Petersen, 1998; Poldrack et al., 1999; Thompson-Schill et al., 1997, 1999).

The Age of Acquisition (AoA) effect also extended to the visual areas, encompassing the bilateral inferior occipital cortices and left occipitotemporal cortex. This suggests that its influence reaches back to the initial stages of object recognition, indicating that objects learned earlier are recognized more efficiently (Catling et al.,

2008; Dent et al., 2007). Specifically, the AoA effect in the left occipitotemporal cortex aligns with the findings of Ellis et al.'s (2006) fMRI study, which attributed this effect to the increased demand for linking visual and semantic representations for items learned later. Yet, considering recent imaging research on the role of this area in picture naming (Mano et al., 2012), it remains plausible that it also processes phonological information. Consequently, our study demonstrates that the AoA effect spans multiple levels of information processing in object naming, including phonological encoding, lexical semantic control, and perceptual processing for object recognition. Our results thus reinforce the conclusions of prior imaging studies regarding the AoA effect (de Zubicaray et al., 2012; Ellis et al., 2006; Fiebach et al., 2003).

9.5.2 The Word Frequency Effect in the Brain

While Morrison and Ellis (1995) initially posited that the word frequency effect was merely an artifact of the age at which words were acquired, subsequent in-depth analyses have distinguished the AoA and word frequency effects as stemming from distinct sources (Belke et al., 2005; Brysbaert, 1996; Brysbaert & Ghyselinck, 2006; Ghyselinck et al., 2004; Perez, 2007). Specifically, word frequency has been shown to primarily influence lexical access during picture naming (Almeida et al., 2007; Cuetos et al., 2010; Jescheniak & Levelt, 1994). Our findings reveal an association between the word frequency effect and activation in the left inferior frontal gyrus, underscoring its involvement in lexical access for word production—a conclusion that aligns with previous research (Grande et al., 2011; Graves et al., 2007). Additionally, activity in the Supplementary Motor Area (SMA), left inferior parietal cortex, and thalamus suggests an increased demand for encoding the motor plan for articulation of low-frequency words, consistent with speech production literature (Ackermann & Riecker, 2004; Bohland & Guenther, 2006).

Notably, our study also highlighted the involvement of the left occipitotemporal cortex, which demonstrated a word frequency effect in picture naming. This aligns with recent findings showing preferential activation in subregions of the left occipitotemporal cortex for reading and object naming tasks, attributed to the demands of mapping visual inputs onto their phonological representations (Mano et al., 2012). This suggests that varying the word frequency of the names for pictured objects impacts this visual-to-phonological mapping process.

9.5.3 Overlap of the AoA and Word Frequency Effects in the Brain

Figure 9.2 reveals that while certain brain regions are uniquely associated with either the AoA or word frequency effect, there is also significant overlap in areas including the inferior frontal cortex, the SMA, and the left occipitotemporal cortex. This overlap suggests that both phonological encoding and the mapping of visual input to its phonological representation are influenced by AoA and word frequency during object naming tasks. The shared involvement of the left occipitotemporal cortex in processing both AoA and word frequency effects underscores its critical role in object naming. Our findings, alongside previous research (Mano et al., 2012), suggest that this area predominantly facilitates the mapping of visual input to phonological output. The concurrent activation related to AoA effects further diminishes the likelihood of its primary role in semantic mapping, a theory previously proposed (Ellis et al., 2006).

9.5.4 Neural Effects of Word Frequency for the Early AoA and Late AoA

The motivation for the two planned contrasts stemmed from prior simulation and behavioral studies suggesting an interaction between word frequency and AoA (Cuetos et al., 2006; Ellis & Lambon Ralph, 2000). Our findings revealed distinct associations: firstly, the ventral part of the inferior frontal gyrus showed a word frequency effect for early learned items (as illustrated in Fig. 9.3A and the yellow portion of Fig. 9.3C). Given the known role of this area in lexical semantic control (Bookheimer, 2002; Fiez & Petersen, 1998; Kuo et al., 2004; Poldrack et al., 1999), this suggests that word frequency influences the semantic processing of early AoA items more significantly. Secondly, the dorsal part of the inferior frontal gyrus and the inferior parietal cortex were linked to the word frequency effect for late-learned items (shown in Fig. 9.3B and the blue portion of Fig. 9.3C). Since these areas have been associated with phonological processing (Bookheimer, 2002; Fiez & Petersen, 1998; Poldrack et al., 1999), it indicates that word frequency impacts the phonological aspect of processing for late AoA items. Previous behavioral studies have highlighted the lexical frequency effect as a key feature of lexical access in picture naming (Almeida et al., 2007; Jescheniak & Levelt, 1994). Our results advance the understanding that lexical access involves two processing aspects—semantic and phonological—that are influenced by the timing of concept acquisition in one's life.

Ellis and Lambon Ralph (2000) discovered through simulation data that both the timing of item introduction into the network (Age of Acquisition, AoA) and the frequency of exposure during training significantly impacted network performance. They found that the network's plasticity decreased as the intermediate layer's state was altered through the early learning of items, leading to the AoA effect. Conversely, items that were trained more frequently showed better recall. Our study, which looked into the neural underpinnings of word frequency effects for items learned early and late in the AoA spectrum, confirms that there is indeed an interaction between these two factors. Furthermore, our results illuminate the path for future research into the mechanisms underlying these effects.

9.6 Conclusion

Research has shown that cortical representations of objects' perceptual, motor, and conceptual attributes are organized in a manner that supports cognitive functions crucial for everyday life, particularly language processing (Bookheimer, 2002; Mahon & Caramazza, 2009; Martin et al., 1996). Reading is a task, which would interact with and influence this organization during development. Our study extends this understanding by exploring how the orchestration between conceptual content and reading processes is facilitated. Overlapping effects in the inferior frontal cortex, SMA, and left occipitotemporal cortex indicate common processes influenced by either AoA or word frequency. Additionally, we identified distinct areas implicated in these effects: the middle temporal gyri's involvement in the AoA effect highlights its role in lexical semantic processing or retrieval, while the participation of subcortical structures and the left inferior parietal cortex in the word frequency effect suggests a primary impact on lexical phonology construction. The discovery of distinct neural networks activated by word frequency effects in early and late AoA further elucidates the dynamic orchestration of conceptual content and process interactions shaped over the course of development.

Acknowledgement This study was supported by the grants from the National Science and Technology Council of Taiwan (112-2410-H-A49-076-MY2; 111-2923-H-A49 002-MY3; 110-2410-H A49A503-MY2).

References

Ackermann, H., & Riecker, A. (2004). The contribution of the insula to motor aspects of speech production: A review and a hypothesis. *Brain and Language, 89*, 320–328.

Almeida, J., Knobel, M., Finkbeiner, M., & Caramazza, A. (2007). The locus of the frequency effect in picture naming: When recognizing is not enough. *Psychonomic Bulletin & Review, 14*, 1177–1182.

Balota, D. A., Cortese, M. J., Sergent-Marshall, S. D., Spieler, D. H., & Yap, M. (2004). Visual word recognition of single-syllable words. *Journal of Experimental Psychology: General, 133*, 283–316.

Belke, E., Brysbaert, M., Meyer, A. S., & Ghyselinck, M. (2005). Age of acquisition effects in picture naming: Evidence for a lexical-semantic competition hypothesis. *Cognition, 96*, B45–54.

Bohland, J. W., & Guenther, F. H. (2006). An fMRI investigation of syllable sequence production. *NeuroImage, 32*, 821–841.

Bookheimer, S. (2002). Functional MRI of language: New approaches to understanding the cortical organization of semantic processing. *Annual Review of Neuroscience, 25*, 151–188.

Brown, G. D., & Watson, F. L. (1987). First in, first out: Word learning age and spoken word frequency as predictors of word familiarity and word naming latency. *Memory and Cognition, 15*, 208–216.

Brysbaert, M. (1996). Word frequency affects naming latency in Dutch when age of acquisition is controlled. *European Journal of Cognitive Psychology, 8*, 185–193.

Brysbaert, M., & Ghyselinck, M. (2006). The effect of age of acquisition: Partly frequency related, partly frequency independent. *Visual Cognition, 13*, 992–1011.

Brysbaert, M., Lange, M., & VAn Wijnendaele, I. (2000). The effects of age-of-acquisition and frequency-of-occurrence in visual word recognition: Further evidence from the Dutch language. *European Journal of Cognitive Psychology, 12*, 65–85.

Carroll, J. B., & White, M. N. (1973). Age-of-acquisition norms for 220 picturable nouns. *Journal of Verbal Learning and Verbal Behavior, 12*, 563–576.

Catling, J. C., Dent, K., & Williamson, S. (2008). Age of acquisition, not word frequency affects object recognition: Evidence from the effects of visual degradation. *Acta Pathologica, Microbiologica, Et Immunologica Scandinavica, 129*, 130–137.

Catling, J. C., & Johnston, R. A. (2009). The varying effects of age of acquisition. *Q J Exp Psychol (Colchester), 62*, 50–62.

Cuetos, F., Alvarez, B., Gonzalez-Nosti, M., Meot, A., & Bonin, P. (2006). Determinants of lexical access in speech production: Role of word frequency and age of acquisition. *Memory & Cognition, 34*, 999–1010.

Cuetos, F., Bonin, P., Alameda, J. R., & Caramazza, A. (2010). The specific-word frequency effect in speech production: Evidence from Spanish and French. *Q J Exp Psychol (Hove), 63*, 750–771.

de Zubicaray, G. I., Miozzo, M., Johnson, K., Schiller, N. O., & McMahon, K. L. (2012). Independent distractor frequency and age-of-acquisition effects in picture-word interference: FMRI evidence for post-lexical and lexical accounts according to distractor type. *Journal of Cognitive Neuroscience, 24*, 482–495.

Dent, K., Catling, J. C., & Johnston, R. A. (2007). Age of acquisition affects object recognition: Evidence from visual duration thresholds. *Acta Pathologica, Microbiologica, Et Immunologica Scandinavica, 125*, 301–318.

Ellis, A. W., Burani, C., Izura, C., Bromiley, A., & Venneri, A. (2006). Traces of vocabulary acquisition in the brain: Evidence from covert object naming. *NeuroImage, 33*, 958–968.

Ellis, A. W., & Lambon Ralph, M. A. (2000). Age of acquisition effects in adult lexical processing reflect loss of plasticity in maturing systems: Insights from connectionist networks. *Journal of Experimental Psychology. Learning, Memory, and Cognition, 26*, 1103–1123.

Fiebach, C. J., Friederici, A. D., Müller, K., & von Cramon, D. Y. (2002). FMRI evidence for dual routes to the mental lexicon in visual word recognition. *Journal of Cognitive Neuroscience, 14*, 11–23.

Fiebach, C. J., Friederici, A. D., Muller, K., von Cramon, D. Y., & Hernandez, A. E. (2003). Distinct brain representations for early and late learned words. *NeuroImage, 19*, 1627–1637.

Fiez, J. A., & Petersen, S. E. (1998). Neuroimaging studies of word reading. *Proc Natl Acad Sci U S A, 95*, 914–921.

Gerhand, S., & Barry, C. (1999). Age-of-acquisition and frequency effects in speeded word naming. *Cognition, 73*, B27-36.

Ghyselinck, M., Lewis, M. B., & Brysbaert, M. (2004). Age of acquisition and the cumulative-frequency hypothesis: A review of the literature and a new multi-task investigation. *Acta Pathologica, Microbiologica, Et Immunologica Scandinavica, 115*, 43–67.

Grande, M., Meffert, E., Huber, W., Amunts, K., & Heim, S. (2011). Word frequency effects in the left IFG in dyslexic and normally reading children during picture naming and reading. *NeuroImage, 57*, 1212–1220.

Graves, W. W., Grabowski, T. J., Mehta, S., & Gordon, J. K. (2007). A neural signature of phonological access: Distinguishing the effects of word frequency from familiarity and length in overt picture naming. *Journal of Cognitive Neuroscience, 19*, 617–631.

Gronholm, P., Rinne, J. O., Vorobyev, V., & Laine, M. (2005). Naming of newly learned objects: A PET activation study. *Brain Research. Cognitive Brain Research, 25*, 359–371.

Hernandez, A., & Fiebach, C. (2006). The brain bases of reading late learned words: Evidence from functional MRI. *Visual Cognition, 13*, 1027–1043.

Hernandez, A. E., & Li, P. (2007). Age of acquisition: Its neural and computational mechanisms. *Psychological Bulletin, 133*, 638–650.

Jescheniak, J. D., & Levelt, W. J. M. (1994). Word frequency effects in speech production: Retrieval of syntactic information and of phonological form. *Journal of Experimental Psychology: Learning, Memory, and Cognition., 20*, 824–843.

Juhasz, B. J. (2005). Age-of-acquisition effects in word and picture identification. *Psychological Bulletin, 131*, 684–712.

Kuo, W. J., Yeh, T. C., Lee, J. R., Chen, L. F., Lee, P. L., Chen, S. S., Ho, L. T., Hung, D. L., Tzeng, O. J., & Hsieh, J. C. (2004). Orthographic and phonological processing of Chinese characters: An fMRI study. *NeuroImage, 21*, 1721–1731.

Levelt, W. J. (1999). Models of word production. *Trends in Cognitive Sciences, 3*, 223–232.

Mahon, B. Z., & Caramazza, A. (2009). Concepts and categories: A cognitive neuropsychological perspective. *Annual Review of Psychology, 60*, 27–51.

Mano, Q.R., Humphries, C., Desai, R. H., Seidenberg, M. S., Osmon, D. C., Stengel, B. C., & Binder, J. R. (2012). *The role of left occipitotemporal cortex in reading: Reconciling stimulus, task, and lexicality effects*. Cereb Cortex.

Martin, A., Wiggs, C. L., Ungerleider, L. G., & Haxby, J. V. (1996). Neural correlates of category-specific knowledge. *Nature, 379*, 649–652.

Morrison, C. M., & Ellis, A. W. (1995). The roles of word frequency and age of acquisition in word naming and lexical decision. *Journal of Experimental Psychology: Learning, Memory, and Cognition, 21*, 116–133.

Morrison, C. M., Ellis, A. W., & Quinlan, P. T. (1992). Age of acquisition, not word frequency, affects object naming, not object recognition. *Memory & Cognition, 20*, 705–714.

Oldfield, R. C. (1971). The assessment and analysis of handedness: The Edinburgh inventory. *Neuropsychologia, 9*, 97–113.

Perez, M. A. (2007). Age of acquisition persists as the main factor in picture naming when cumulative word frequency and frequency trajectory are controlled. *Q J Exp Psychol (Colchester), 60*, 32–42.

Poldrack, R. A., Wagner, A. D., Prull, M. W., Desmond, J. E., Glover, G. H., & Gabrieli, J. D. (1999). Functional specialization for semantic and phonological processing in the left inferior prefrontal cortex. *NeuroImage, 10*, 15–35.

Price, C. J. (2000). The anatomy of language: Contributions from functional neuroimaging. *Journal of Anatomy, 197*(Pt 3), 335–359.

Price, C. J. (2012). A review and synthesis of the first 20years of PET and fMRI studies of heard speech, spoken language and reading. *NeuroImage, 62*, 816–847.

Szekely, A., D'Amico, S., Devescovi, A., Federmeier, K., Herron, D., Iyer, G., Jacobsen, T., & Bates, E. (2003). Timed picture naming: Extended norms and validation against previous studies. *Behavior Research Methods, Instruments, & Computers, 35*, 621–633.

Thompson-Schill, S. L., D'Esposito, M., Aguirre, G. K., & Farah, M. J. (1997). Role of left inferior prefrontal cortex in retrieval of semantic knowledge: A reevaluation. *Proc Natl Acad Sci U S A, 94*, 14792–14797.

Thompson-Schill, S. L., D'Esposito, M., & Kan, I. P. (1999). Effects of repetition and competition on activity in left prefrontal cortex during word generation. *Neuron, 23*, 513–522.

van Turennout, M., Bielamowicz, L., & Martin, A. (2003). Modulation of neural activity during object naming: Effects of time and practice. *Cerebral Cortex, 13*, 381–391.

van Turennout, M., Ellmore, T., & Martin, A. (2000). Long-lasting cortical plasticity in the object naming system. *Nature Neuroscience, 3*, 1329–1334.

Zevin, J. D., & Seidenberg, M. S. (2004). Age-of-acquisition effects in reading aloud: Tests of cumulative frequency and frequency trajectory. *Memory & Cognition, 32*, 31–38.

Part IV
Brain Computations of Language Processing and Beyond

Chapter 10
Language Comprehension, Chinese Reading, and the Aging Brain

Nannan Zhou, Chih-Mao Huang, and Hsu-Wen Huang

Abstract Age-related changes in cognitive and language functions have been extensively studied in the last 50 years. Older adults represent a unique population for the study of language because of the many challenges associated with studying this population, including individual differences in life experiences, neurocognitive functions, and social factors. Cross-sectional functional neuroimaging studies have shown that older and younger adults engage similar language-associated brain regions, suggesting that the general architecture of the language network is preserved across the lifespan. However, electrophysiological research has shown that the N400 response involved in semantic processing is reduced and delayed with aging. To date, few studies have examined the aging-related trajectories of Chinese reading and language comprehension. In contrast with English words, most Chinese words are composed of two syllables, which can be mapped onto one or more morphemes with clear boundaries. This chapter briefly reviews the behavioral and neuroimaging research on Chinese word processing throughout the adult lifespan. It explores orthographic, phonological, sublexical, and lexicosemantic processes, considering neural mechanisms and neuropsychological, psycholinguistic, and cognitive reserve factors. This review advances our understanding of the neurocognitive mechanisms underlying language comprehension and Chinese word reading in the aging brain.

Keywords Cognitive aging · Language comprehension · Chinese word processing · Neurocognitive mechanisms · Event-related potentials · Sublexical ambiguity

N. Zhou · H.-W. Huang (✉)
National Center for Geriatrics and Welfare Research, National Health Research Institutes, Miaoli, Taiwan
e-mail: hwhuang@nhri.edu.tw

C.-M. Huang
Department of Biological Science and Technology, National Yang Ming Chiao Tung University, Hsin-Chu, Taiwan

Department of Psychology, The University of Hong Kong, Hong Kong, Hong Kong

D. H. Wu (ed.), *Cognitive and Neural Foundations of Chinese Reading*, Chinese Language Learning Sciences, https://doi.org/10.1007/978-981-96-6669-0_10

10.1 Overview

This chapter explores cognitive aging and its impact on language processing, with a specific focus on word reading and recognition, because words are the basic building blocks of language. The mental ability to comprehend language, a central aspect of human life, involves a complex array of sensory and cognitive processes. These processes, ranging from analyzing sensory inputs to cueing long-term semantic memory, are essential for acquiring information, completing tasks, and maintaining social relationships. While language comprehension often appears seamless, it is underpinned by intricate subprocesses, all of which undergo changes with normal aging.

As we age, our understanding of words and their meanings, which constitutes a core component of our semantic memory, generally remains stable and may even improve in some cases (see review by Light, 1992). However, the real-time language processing of this semantic information, especially at the extremely high rates required for speech, is where the effects of aging become more apparent (Federmeier et al., 2003; Huang C.M. et al., 2012). Sensory thresholds increase with age, affecting the initial analysis of sensory streams. This may lead to delays or variability in word recognition and lexical access.

The aging brain exhibits an increased dependency on contextual information for language processing. This phenomenon is evident in larger behavioral semantic priming effects and delayed N400 responses in older adults, indicating a shift in how semantic information is accessed and integrated (Kutas & Iragui, 1998). These changes in the N400 response, a well-established electrophysiological component linked to semantic processing, underscore the nuanced ways in which aging impacts the brain's ability to process language.

Focusing on Chinese reading, we examine the unique challenges presented by its structure, which is significantly different from alphabetic languages. The complexity of Chinese words, typically composed of two syllables and multiple morphemes, offers a distinct perspective on the effects of aging on language processing (Huang & Huang, 2022). This chapter unravels the impact of aging on various aspects of information processing in reading, particularly in the context of the Chinese language, considering the cognitive, sensory, and semantic changes that occur with aging.

10.2 Cognitive Aging and Its Impact on Word Reading and Recognition

Studies examining the impact of aging on cognitive function, particularly in language processing, have revealed a narrative characterized by varying degrees of cognitive resilience and decline. Burke and Peters (1986) and Burke et al. (1987) emphasized the preservation of semantic encoding and priming abilities in older adults, indicating cognitive resilience (Park et al., 2002). Cohen-Shikora and Balota (2016) observed consistent effects of lexical variables, such as length, word frequency, and orthographic neighborhood, on visual word recognition throughout the adult lifespan, but a decline in recognition speed among older adults. These findings can be attributed to the complex interplay of factors in the cognitive dynamics associated with aging, including age-related increases in verbal knowledge and declines in processing speed, vision, working memory, and executive functions (Park et al., 2002; Froehlich et al., 2016; Huang & Huang, 2019).

Several studies have reported structural and functional brain changes associated with aging; these changes influence cognitive compensation, maintenance, and reserve (Cabeza et al., 2018; Park & Reuter-Lorenz, 2009). These factors contribute to the outcomes of healthy aging or dementia, impacting complex tasks such as language processing.

Understanding the atrophy of the aging brain is essential to understand age-related changes and individual differences in cognitive and language functions. A wealth of evidence across several cross-sectional and longitudinal studies demonstrates that biological aging is characterized by reductions in gray matter volume (Raz et al., 2005; Resnick et al., 2003; Rodrigue & Raz, 2004) and cortical thickness (Dickerson et al., 2009; Fjell et al., 2009). The prefrontal, parietal, and hippocampal regions show the greatest shrinkage, whereas the temporal and occipital regions show minimal decreases. Although the associations between structure and cognition are not easily replicated and appear to be sensitive to subtypes of neurocognitive assessment and sample composition, studies have indicated that these age-related reductions in brain structure may index specific cognitive decline in older adults. For example, age-related volumetric shrinkage in the prefrontal cortex has been found to be associated with lower fluid intelligence (Fjell et al., 2006; Head et al., 2008), poorer inhibitory/motor control and working memory performance (Head et al., 2009; Tapp et al., 2004), and enhanced verbal fluency (Elderkin-Thompson et al., 2008). Moreover, biological aging is associated with reductions in white matter volume, density, and fiber integrity (Cox et al., 2016; Fan et al., 2019; Walhovd et al., 2011). The most consistent age-related change in white matter fiber integrity is a greater reduction in the anterior regions of the corpus callosum, especially the anterior regions and body, relative to the posterior regions of the corpus callosum (Davis et al., 2012; Gordon et al., 2008; Sullivan et al., 2010). This points to an anterior-to-posterior deterioration in white-matter diffusivity and anisotropy with age. Fan et al. (2019) linked cognitive decline with deterioration in white matter integrity, observing age-related changes in white matter integrity in the anterior portion of the corpus callosum, the right uncinate fasciculus, and the cingulum-angular fibers. Their findings demonstrated that increased fractional anisotropy in the anterior corpus callosum and right cingulum-angular fibers was positively correlated with enhanced performance in visuospatial tasks in older adults. This finding highlights the vulnerability of frontal white matter fibers to aging, indicating the mediation effect of white matter integrity on the relationship between aging and visuospatial performance in older adults.

Age-related alterations in brain activity patterns have been consistently observed across diverse cognitive and motor domains, including language comprehension, working memory, cognitive control, and motor function (Cabeza et al., 2018; Grady,

2012; Park & Reuter-Lorenz, 2009). Using noninvasive functional neuroimaging techniques, such as functional magnetic resonance imaging, studies have identified two predominant patterns in healthy older individuals. First, older individuals often exhibit greater bilateral activity than their younger counterparts, who display a higher level of unilateral activity. Second, older individuals may show additional and more distributed activation in regions that are typically not activated in young adults (Cabeza et al., 2002, 2004; Jimura & Braver, 2010; Huang C.M. et al., 2012; Huang H.W. et al., 2012). The increased activity observed in frontoparietal regions with aging is commonly interpreted as a compensatory mechanism that plays a role in maintaining or improving behavioral performance in response to age-related neurodegeneration (Cabeza et al., 2002; Davis et al., 2008; Heuninckx et al., 2008; Vallesi et al., 2011; Huang C.M. et al., 2012; Huang H.W. et al., 2012). This phenomenon reflects the adaptive nature of the aging brain, which uses increased neural engagement to mitigate the effects of age-related cognitive declines. Furthermore, converging evidence shows that aging processes are associated with significant changes in functional connectivity between different brain regions, resulting in less effective neural communication in the aging brain (Andrews-Hanna et al., 2007). Studies measuring functional connectivity during the resting state in predefined brain regions have demonstrated that intrinsic brain connectivity in the aging brain is related to performance in diverse cognitive tasks, such as executive function and processing speed (Damoiseaux et al., 2008), associative memory (Wang et al., 2010), and working memory (Sambataro et al., 2010). More recently, Patil et al. (2021) revealed age-related alterations in functional connectivity related to creative cognition within the brain's default mode network (DMN) and cerebellar network. Compared with young adults, older individuals exhibited less specific and more constrained connectivity between the DMN, fronto-parietal executive control network, and visual, auditory, and cerebellar networks related to creativity, which can affect language use.

Froehlich et al. (2018) examined the impact of aging on the brain and reading and found that the overall architecture and neural circuits associated with reading were preserved across the lifespan. However, they observed age-related variations in brain activity within specific subcomponents of the reading network, such as the phonological and orthographic subprocesses. These differences may result from older adults' inability to inhibit less effective reading strategies.

To counteract age-related neural changes, older adults deploy adaptive word processing strategies. In addition, certain neurocognitive reserve factors play a role in mitigating age-related declines in language processing abilities among older individuals. Froehlich et al. (2016) observed that older adults displayed longer response times and required more information to complete language tasks. Adapting to this high information requirement, high-performing older readers even outperformed younger adults in the speed of information uptake during orthographic and lexicosemantic processing. Steen-Baker et al. (2017) emphasized the protective effect of literacy against age-related declines in language processing. They found an interaction between age and literacy skills: participants with higher literacy skills showed a reduction in the duration of first fixation on words as they aged, whereas those with lower literacy skills exhibited an increase in this duration. This pattern suggests that literacy skills act as a buffer against age-related decline in reading. In addition to language-related factors, Stern (2009) suggested that higher education levels, greater occupational complexity, and engagement in mentally stimulating activities can enhance older adults' ability to cope effectively with various age-related cognitive and brain changes, emphasizing the role of neurocognitive reserve. Similarly, Zhou et al. (2022) and Li et al. (2025) investigated the impact of social and emotional factors on language processing in older adults. They discovered that lower perceived loneliness was correlated with a superior ability to resolve lexical ambiguity and/or to retrieve word meaning, underscoring the role of social contact in mitigating language processing challenges in older adults. Collectively, these studies depict the influence of aging on language processing as a dynamic interplay between declining neural infrastructure and the implementation of adaptive cognitive strategies.

The exploration of the intricate ways in which cognitive aging impacts language processing, particularly in the context of word processing, indicates that this influence extends beyond mere cognitive mechanisms. In the following section, we delve deeper into how aging specifically affects different dimensions of word processing, including orthographic, phonological, and semantic dimensions. This exploration provides a more comprehensive understanding of the nuanced changes that occur in these key areas of language processing in older adults.

10.3 Age-Related Changes in Word Processing: Orthographic, Phonological, and Semantic Perspectives

Studies examining age-related alterations in word processing have revealed distinct patterns across various dimensions, including phonological, orthographic, and semantic dimensions. Phonological and orthographic processing typically declines with aging, as evidenced by a reduced efficiency in temporal sampling among older adults (Froehlich et al., 2016). Federmeier et al. (2003) observed delayed N1 and P2 components in older adults when exposed to natural speech, reflecting delayed sensory and attention-related neural responses. In another study by Froehlich et al. (2018), orthographic processing was assessed using a lexical decision task. The participants were presented with either real words or pseudowords. The pseudowords were specifically designed to be pronounceable and share the same pronunciation as existing real words. The participants were tasked with deciding whether each presented item was a real word while potentially encountering a challenge posed by pseudowords that mimicked real-word pronunciations. Conversely, phonological processing was assessed through a task requiring the participants to judge whether a presented pseudoword shared the same pronunciation as a real word. The two tasks involved distinct cognitive processes and required the participants to use different strategies. In the lexical decision task, which assessed orthographic processing, the participants were required to focus on word forms and make judgments based on spelling. Conversely, the phonological decision task involved assessing the pronunciation of presented items, requiring participants to make judgments based on sound. The challenge arose from the need to inhibit one processing strategy while deploying the other. Notably, older individuals showed activity in brain regions associated with both strategies, indicating difficulties with their inhibitory mechanisms.

Whereas phonological and orthographic subprocesses typically decline with age, semantic processing demonstrates a mixed pattern of both resilience and decline in older adults. Older adults demonstrate semantic priming and neighborhood size effects similar to those of younger adults, observable across both behavioral and neurobiological measures; this indicates the preservation of the organizational structure of the semantic network during normal aging (Burke et al., 1987; Federmeier et al., 2003; Jongman & Federmeier, 2022). However, subtle yet robust age-related changes in brain activation are evident in event-related potential (ERP) results. The N400 difference, which reflects contextual congruency effects, is both smaller and delayed in older adults, indicating a diminished ability to use contextual information for guiding semantic processing. This suggests that aging not only impacts initial sensory processing but also hinders higher-order language processes (Federmeier & Kutas, 2005; Federmeier et al., 2003). Older adults exhibit a diminished capacity to adjust activation in the semantic network, as indicated by a weaker repetition effect on the N400 component compared with that in their younger counterparts (Jongman & Federmeier, 2022). Moreover, older adults struggle to revise initial interpretations of ambiguous words when presented with conflicting information (Meyer & Federmeier, 2010). Furthermore, the inhibitory effect of neighborhood frequency observed in young adults is absent in older adults during lexical decision tasks, indicating an age-related decline in lexical inhibition and activation (Robert & Mathey, 2007).

Older adults' semantic processing, compared with that of their younger counterparts, is characterized by increased passivity, diminished automaticity, and reduced consistency in prediction (Federmeier et al., 2010; Jongman et al., 2023). Huang H.W. et al. (2012) explored this aspect further by presenting participants with adjective–noun phrases in which the adjectives conveyed either concrete or abstract meanings. Although older adults displayed a concreteness effect at the adjectives similar to that of younger participants, this effect did not extend to the nouns. In contrast with younger individuals, older adults did not exhibit concreteness-based predictability effects associated with left hemisphere processing or imagery effects linked to right hemisphere processing at the noun. This observation suggests that although verbal knowledge remains relatively stable across the lifespan, age-related changes manifest in the automatic use of that knowledge during online language processing. Supporting this notion, another study found that older adults exhibited left-lateralized frontal negativity for predictable words at approximately 500 ms after stimulus presentation, indicating a lack of anticipation and the need to revisit contextual material to form an integrated meaning (Wlotko et al., 2012). Furthermore, the asymmetric reduction of the N400 response to expected words, which is typically observed in younger adults, with a more pronounced reduction in the left hemisphere, is not noted in older adults. This discrepancy suggests a shift in older adults toward more distributed and bilateral neural activities during semantic processing (Federmeier & Kutas, 2019).

In summary, aging predominantly influences the speed and efficiency of orthographic and phonological processing; semantic processing demonstrates a degree of resilience to aging. However, discernible changes in strategies and neural mechanisms used by older adults in semantic processing emerge from neurobiological studies.

Having examined age-related changes in orthographic, phonological, and semantic word processing, we now focus on Chinese word processing. The next section explores how these general patterns of aging in language processing manifest within the unique structure and characteristics of the Chinese language, offering insights into the complex interaction among aging, language structure, and cognitive functions.

10.4 The Complexity of Chinese Word Processing

The intricate relationship between words and their meanings introduces complexity through the phenomenon of lexical ambiguity, where a single word can have multiple meanings. Lexical ambiguity is categorized into two types: homonymy and polysemy. Homonymous words, such as "bark," have distinct and unrelated meanings for a single word form, such as the sound made by a dog and the outer layer of a tree. Polysemous words, such as "paper," have semantically related senses associated with a single word form; for example, "paper" refers to both a material and a newspaper publication.

In the domain of visual word recognition, early research consistently reported a phenomenon termed the "ambiguity advantage effect," wherein words with multiple meanings yield faster response times than those with fewer meanings, even when matched for frequency (Azuma & Van Orden, 1997; Borowsky & Masson, 1996; Hino & Lupker, 1996; Jastrzembski, 1981; Kellas et al., 1988; Lichacz et al., 1999; Millis & Button, 1989; Rubenstein et al., 1970). This effect has traditionally been attributed to the processing benefit derived from multiple entries in the lexicon. However, recent challenges to this assumption have indicated the need for a nuanced understanding of ambiguity, distinguishing between unrelated meanings (homonymy) and related senses (polysemy).

To assess the impact of unrelated meanings and related senses, Rodd et al. (2002) reanalyzed studies using high- and low-ambiguity words (e.g., Azuma & Van Orden, 1997; Borowsky & Masson, 1996; Millis & Button, 1989). Contrary to expectations, they found that high-ambiguity words did not differ in the number of meanings from low-ambiguity words; instead, high-ambiguity words exhibited a significantly higher number of related senses. This led to the proposition that the ambiguity advantage is linked to polysemous words with numerous related senses. In their lexical decision task, Rodd et al. (2002) identified opposing effects of homonymy and polysemy, with a processing disadvantage for words with multiple unrelated meanings (homonyms) and an advantage for words with numerous interrelated senses (polysemes). Subsequently, these findings were supported by studies highlighting the representational and processing differences between homonymy and polysemy (Frazier & Rayner, 1990; Klepousniotou, 2002; Klepousniotou et al., 2012).

In addition to lexical decision tasks, electroencephalography (EEG) and magnetoencephalography (MEG) studies have examined brain responses during word reading. MEG revealed differential patterns in the M350 component, which is sensitive to lexical activation, depending on lexical ambiguities. Homonyms exhibited delayed M350 latency, whereas polysemous words with multiple senses showed an earlier peak latency, further indicating the presence of neurocognitive disparities between homonymy and polysemy (Beretta et al., 2005; Pylkkänen et al., 2006).

These insights have predominantly been derived from alphabetic writing systems, such as English. A crucial question arises regarding their applicability to Chinese, which is primarily characterized by a disyllabic compound word structure. Chinese compounds present a unique challenge because they involve characters with distinguishable meanings. Resolving lexical ambiguity in Chinese becomes more intricate due to the potential ambiguity of characters at the sublexical level and their contribution to whole-word meaning. To address this challenge, Huang H.W. et al. (2011) manipulated subject-rated sublexical ambiguity in a lexical decision task. Their findings revealed that words with high sublexical semantic ambiguity elicited smaller N400 amplitudes than those with low ambiguity. However, this observation contradicts the findings of previous ambiguity studies (e.g., Beretta et al., 2005), prompting a reconsideration of the concepts of homonymy and polysemy, particularly at the sublexical level, in the context of Chinese word processing.

Building upon this notion, Huang C.Y. et al. (2011) manipulated the visual field and the number of senses of the first character in Chinese disyllabic compounds to investigate the representation and processing of related senses in the two brain hemispheres. ERP results revealed a sense facilitation effect in the left hemisphere (LH) during less demanding lexical processing tasks and in the right hemisphere (RH) during more demanding lexical processing tasks. These findings align with the hypothesis of a single-entry representation of senses and the fine/coarse coding hypothesis in the LH/RH, as proposed by Beeman (1998) and Bouaffre and Fa ta-Ainseba (2007).

To further examine the potential interaction between meanings and senses, Huang and Lee (2018) manipulated the number of meanings (NOM) and the number of senses (NOS) of the first constituent characters in Chinese compound words. Their investigation revealed interactions between NOM and NOS in both behavioral results and N400 responses. Specifically, the NOM disadvantage effect was evident only for words with few senses, whereas the NOS facilitation effect was significant only for words with multiple meanings. These findings support the single/multiple entry hypothesis for the senses/meanings of constituent characters in Chinese compound words.

Studies on lexical ambiguity in word processing have predominantly focused on younger adults. Given the unique structural complexities and demands of the Chinese language, there is a critical need for more focused research on how aging affects

Chinese word processing. This area remains largely unexplored, and the distinct characteristics of Chinese call for an extension of these studies to include aging populations.

Helping to answer this call, Zhou et al. (2022) leveraged the NOM disadvantage effect to examine the impact of aging and individual differences on ambiguity resolution in older adults. They identified an increased lexicality effect (where real words are identified more easily than pseudowords) and a comparable sublexical ambiguity effect (where sublexical unambiguous words were identified more easily than sublexical ambiguous words) in older adults than in their younger counterparts. This finding suggests that aging does not uniformly influence all aspects of word processing. Although their ability to resolve sublexical ambiguities remained similar to that of their younger counterparts, older adults exhibited perceived loneliness as a significant risk factor for ambiguity resolution, emphasizing the intricate interplay between social connections and language processing in aging.

In summary, the exploration of lexical ambiguity in word processing has revealed clear differences between homonymy and polysemy, pointing to distinct behavioral and neurocognitive differences in the processing of unrelated meanings and related senses. Although early studies focused on alphabetic writing systems, the transition to Chinese compounds highlighted the significance of sublexical ambiguity, refining existing models, such as the single- and multiple-entry representation of senses and meanings. Extending this line of inquiry to aging populations has uncovered associations among language processing, ambiguity resolution, and social factors. Future research holds the potential to unravel even more intricate patterns and provide a comprehensive understanding of how language processing adapts and evolves across different linguistic and demographic contexts.

10.5 Conclusion

Concluding this exploration of cognitive aging and its impact on language processing, particularly Chinese word reading, it is evident that aging brings about a complex interplay of decline and resilience in various aspects of language processing. Despite declines in sensory thresholds and potentially in word recognition and lexical access, there is a notable stability or even enhancement in the knowledge of words and meanings among older adults. This duality highlights the adaptive nature of the aging brain in the realm of language comprehension.

Due to the unique linguistic structure of Chinese, focusing on Chinese reading offers a deeper understanding of how aging influences language processing. The intricate relationship between Chinese characters and their meanings presents both challenges and opportunities for older adults in engaging with language.

The interplay between social connections and language processing in older adults is becoming increasingly relevant, with research suggesting that factors such as literacy, social engagement, and cognitive reserve play pivotal roles in how older adults navigate and compensate for the changes in language processing due to aging.

This understanding opens avenues for developing strategies and interventions that support language functions in older adults, enhancing their cognitive well-being and overall quality of life.

In summary, the study of language comprehension and Chinese word reading in the aging brain provides unique insights into how aging affects language processing. While certain aspects of language processing decline with age, there is also evidence of remarkable cognitive resilience and adaptability. This understanding not only enriches our knowledge in the fields of cognitive neuroscience and psycholinguistics but also has profound implications for supporting an aging population in our increasingly linguistically diverse societies.

Acknowledgement This work was supported by the National Health Research Institutes, Taiwan (CG-113-GP-15). The authors would like to thank Dr. Bing Li for her assistance in drafting this chapter. Professional English language editing support provided by AsiaEdit.

References

Andrews-Hanna, J. R., Snyder, A. Z., Vincent, J. L., Lustig, C., Head, D., Raichle, M. E., & Buckner, R. L. (2007). Disruption of large-scale brain systems in advanced aging. *Neuron, 56*(5), 924–935.

Azuma, T., & Van Orden, G. C. (1997). Why SAFE is better than FAST: The relatedness of a word's meanings affects lexical decision times. *Journal of Memory and Language, 36*(4), 484–504.

Beeman, M. (1998). Coarse semantic coding and discourse comprehension. In M. Beeman & C. Chiarello (Eds.), *Right hemisphere language comprehension: Perspectives from cognitive neuroscience* (pp. 255–284). Lawrence Erlbaum Associates Publishers.

Beretta, A., Fiorentino, R., & Poeppel, D. (2005). The effects of homonymy and polysemy on lexical access: An MEG study. *Cognitive Brain Research, 24*(1), 57–65.

Borowsky, R., & Masson, M. E. (1996). Semantic ambiguity effects in word identification. *Journal of Experimental Psychology: Learning, Memory, and Cognition, 22*(1), 63.

Bouaffre, S., & Faita-Ainseba, F. (2007). Hemispheric differences in the time-course of semantic priming processes: Evidence from event-related potentials (ERPs). *Brain and Cognition, 63*(2), 123–135.

Burke, D. M., & Peters, L. (1986). Word associations in old age: Evidence for consistency in semantic encoding during adulthood. *Psychology and Aging, 1*(4), 283–292.

Burke, D. M., White, H., & Diaz, D. L. (1987). Semantic priming in young and older adults: Evidence for age constancy in automatic and attentional processes. *Journal of Experimental Psychology: Human Perception and Performance, 13*(1), 79–88.

Cabeza, R., Albert, M., Belleville, S., Craik, F. I. M., Duarte, A., Grady, C. L., Lindenberger, U., Nyberg, L., Park, D. C., Reuter-Lorenz, P. A., Rugg, M. D., Steffener, J., & Rajah, M. N. (2018). Maintenance, reserve and compensation: The cognitive neuroscience of healthy ageing. *Nature Reviews Neuroscience, 19*(11), 701–710.

Cabeza, R., Anderson, N. D., Locantore, J. K., & McIntosh, A. R. (2002). Aging gracefully: Compensatory brain activity in high-performing older adults. *NeuroImage, 17*(3), 1394–1402.

Cabeza, R., Daselaar, S. M., Dolcos, F., Prince, S. E., Budde, M., & Nyberg, L. (2004). Task-independent and task-specific age effects on brain activity during working memory, visual attention and episodic retrieval. *Cerebral Cortex, 14*(4), 364–375.

Cohen-Shikora, E. R., & Balota, D. A. (2016). Visual word recognition across the adult lifespan. *Psychology and Aging, 31*(5), 488–502.

Cox, S. R., Ritchie, S. J., Tucker-Drob, E. M., Liewald, D. C., Hagenaars, S. P., Davies, G., Wardlaw, J. M., Gale, C. R., Bastin, M. E., & Deary, I. J. (2016). Ageing and brain white matter structure in 3,513 UK Biobank participants. *Nature Communications, 7*(1), 13629.

Damoiseaux, J. S., Beckmann, C. F., Arigita, E. J. S., Barkhof, F., Ph., Scheltens, Stam, C. J., Smith, S. M., & Rombouts, S. A. R. B. (2008). Reduced resting-state brain activity in the "default network" in normal aging. *Cerebral Cortex, 18*(8), 1856–1864.

Davis, S. W., Dennis, N. A., Daselaar, S. M., Fleck, M. S., & Cabeza, R. (2008). Que PASA? The posterior–anterior shift in aging. *Cerebral Cortex, 18*(5), 1201–1209.

Davis, S. W., Kragel, J. E., Madden, D. J., & Cabeza, R. (2012). The architecture of cross-hemispheric communication in the aging brain: Linking behavior to functional and structural connectivity. *Cerebral Cortex, 22*(1), 232–242.

Dickerson, B. C., Feczko, E., Augustinack, J. C., Pacheco, J., Morris, J. C., Fischl, B., & Buckner, R. L. (2009). Differential effects of aging and Alzheimer's disease on medial temporal lobe cortical thickness and surface area. *Neurobiology of Aging, 30*(3), 432–440.

Elderkin-Thompson, V., Ballmaier, M., Hellemann, G., Pham, D., & Kumar, A. (2008). Executive function and MRI prefrontal volumes among healthy older adults. *Neuropsychology, 22*(5), 626.

Fan, Y. T., Fang, Y. W., Chen, Y. P., Tzeng, O. J. L., Huang, H. W., & Huang, C. M. (2019). Aging, cognition, and the brain: Effects of age-related variation in white matter integrity on neuropsychological function. *Aging and Mental Health, 23*, 831–839.

Federmeier, K. D., & Kutas, M. (2005). Aging in context: Age-related changes in context use during language comprehension. *Psychophysiology, 42*(2), 133–141.

Federmeier, K. D., & Kutas, M. (2019). What's "left"? Hemispheric sensitivity to predictability and congruity during sentence reading by older adults. *Neuropsychologia, 133*, Article 107173.

Federmeier, K. D., Kutas, M., & Schul, R. (2010). Age-related and individual differences in the use of prediction during language comprehension. *Brain and Language, 115*(3), 149–161.

Federmeier, K. D., Van Petten, C., Schwartz, T. J., & Kutas, M. (2003). Sounds, words, sentences: Age-related changes across levels of language processing. *Psychology and Aging, 18*(4), 858–872.

Fjell, A. M., Walhovd, K. B., Reinvang, I., Lundervold, A., Salat, D., Quinn, B. T., Fischl, B., & Dale, A. M. (2006). Selective increase of cortical thickness in high-performing elderly—Structural indices of optimal cognitive aging. *NeuroImage, 29*(3), 984–994.

Fjell, A. M., Westlye, L. T., Amlien, I., Espeseth, T., Reinvang, I., Raz, N., Agartz, I., Salat, D. H., Greve, D. N., Fischl, B., Dale, A. M., & Walhovd, K. B. (2009). High consistency of regional cortical thinning in aging across multiple samples. *Cerebral Cortex, 19*(9), 2001–2012.

Frazier, L., & Rayner, K. (1990). Taking on semantic commitments: Processing multiple meanings vs. multiple senses. *Journal of Memory and Language, 29*(2), 181–200.

Froehlich, E., Liebig, J., Morawetz, C., Ziegler, J. C., Braun, M., Heekeren, H. R., & Jacobs, A. M. (2018). Same same but different: Processing words in the aging brain. *Neuroscience, 371*, 75–95.

Froehlich, E., Liebig, J., Ziegler, J. C., Braun, M., Lindenberger, U., Heekeren, H. R., & Jacobs, A. M. (2016). Drifting through basic subprocesses of reading: A hierarchical diffusion model analysis of age effects on visual word recognition. *Frontiers in Psychology, 7*, 1863.

Grady, C. (2012). Trends in neurocognitive aging. *Nature Reviews Neuroscience, 13*(7), 491.

Gordon, B. A., Rykhlevskaia, E. I., Brumback, C. R., Lee, Y., Elavsky, S., Konopack, J. F., McAuley, E., Kramer, A. F., Colcombe, S., Gratton, G., & Fabiani, M. (2008). Neuroanatomical correlates of aging, cardiopulmonary fitness level, and education. *Psychophysiology, 45*(5), 825–838.

Head, D., Kennedy, K. M., Rodrigue, K. M., & Raz, N. (2009). Age differences in perseveration: Cognitive and neuroanatomical mediators of performance on the Wisconsin Card Sorting Test. *Neuropsychologia, 47*(4), 1200–1203.

Head, D., Rodrigue, K. M., Kennedy, K. M., & Raz, N. (2008). Neuroanatomical and cognitive mediators of age-related differences in episodic memory. *Neuropsychology, 22*(4), 491–507.

Heuninckx, S., Wenderoth, N., & Swinnen, S. P. (2008). Systems neuroplasticity in the aging brain: Recruiting additional neural resources for successful motor performance in elderly persons. *Journal of Neuroscience, 28*(1), 91–99.

Hino, Y., & Lupker, S. J. (1996). Effects of polysemy in lexical decision and naming: An alternative to lexical access accounts. *Journal of Experimental Psychology: Human Perception and Performance, 22*(6), 1331.

Huang, C. M., & Huang, H. W. (2019). Aging, neurocognitive reserve, and the healthy brain. In *Psychology of learning and motivation* (Vol. 71, pp. 175–213). Elsevier.

Huang, C. M., Polk, T. A., Goh, J. O., & Park, D. C. (2012). Both left and right posterior parietal activations contribute to compensatory processes in normal aging. *Neuropsychologia, 50*, 55–66.

Huang, C. Y., Lee, C. Y., Huang, H. W., & Chou, C. J. (2011). Number of sense effects of Chinese disyllabic compounds in the two hemispheres. *Brain and Language, 119*(2), 99–109.

Huang, H. W., & Huang, C. M. (2022). Developing cross-cultural neuropsychology through the lens of cross-cultural cognitive neuroscience. In A. L. Fernández & J. Evans (Eds.), *Understanding cross-cultural neuropsychology: Science, testing and challenge* (pp. 29–43). Routledge.

Huang, H. W., & Lee, C. Y. (2018). Number of meanings and number of senses: An ERP study of sublexical ambiguities in reading Chinese disyllabic compounds. *Frontiers in Psychology, 9*, 324.

Huang, H. W., Lee, C. Y., Tsai, J. L., & Tzeng, O. J. L. (2011). Sublexical ambiguity effect in reading Chinese disyllabic compounds. *Brain and Language, 117*(2), 77–87.

Huang, H. W., Meyer, A. M., & Federmeier, K. D. (2012). A "concrete view" of aging: Event related potentials reveal age-related changes in basic integrative processes in language. *Neuropsychologia, 50*(1), 26–35.

Jastrzembski, J. E. (1981). Multiple meanings, number of related meanings, frequency of occurrence, and the lexicon. *Cognitive Psychology, 13*(2), 278–305.

Jimura, K., & Braver, T. S. (2010). Age-related shifts in brain activity dynamics during task switching. *Cerebral Cortex, 20*, 1420–1431.

Jongman, S. R., Copeland, A., Xu, Y., Payne, B. R., & Federmeier, K. D. (2023). Older adults show intraindividual variation in the use of predictive processing. *Experimental Aging Research, 49*(5), 433–456.

Jongman, S. R., & Federmeier, K. D. (2022). Age-related changes in the structure and dynamics of the semantic network. *Language, Cognition and Neuroscience, 37*(7), 805–819.

Kellas, G., Ferraro, F. R., & Simpson, G. B. (1988). Lexical ambiguity and the timecourse of attentional allocation in word recognition. *Journal of Experimental Psychology: Human Perception and Performance, 14*(4), 601.

Klepousniotou, E. (2002). The processing of lexical ambiguity: Homonymy and polysemy in the mental lexicon. *Brain and Language, 81*(1–3), 205–223.

Klepousniotou, E., Pike, G. B., Steinhauer, K., & Gracco, V. (2012). Not all ambiguous words are created equal: An EEG investigation of homonymy and polysemy. *Brain and Language, 123*(1), 11–21.

Kutas, M., & Iragui, V. (1998). The N400 in a semantic categorization task across 6 decades. *Electroencephalography and Clinical Neurophysiology, 108*(5), 456–471.

Li, B., Huang, C. M., Wang, Y. Y., Lin, Q., & Huang, H. W. (2025). Loneliness modulates the neural dynamics of semantic processing in healthy older adults: Evidence from event-related potentials. *Social Cognitive and Affective Neuroscience, 20*(1), nsaf030.

Lichacz, F. M., Herdman, C. M., Lefevre, J. A., & Baird, B. (1999). Polysemy effects in word naming. *Canadian Journal of Experimental Psychology/revue Canadienne De Psychologie Expérimentale, 53*(2), 189.

Light, L. L., et al. (1992). The organization of memory in old age. In F. I. M. Craik & T. A. Salthouse (Eds.), *The handbook of aging and cognition* (pp. 111–165). Erlbaum.

Meyer, A. M., & Federmeier, K. D. (2010). Event-related potentials reveal the effects of aging on meaning selection and revision. *Psychophysiology, 47*(4), 673–686.

Millis, M. L., & Button, S. B. (1989). The effect of polysemy on lexical decision time: Now you see it, now you don't. *Memory & Cognition, 17*, 141–147.

Park, D. C., Lautenschlager, G., Hedden, T., Davidson, N. S., Smith, A. D., & Smith, P. K. (2002). Models of visuospatial and verbal memory across the adult life span. *Psychology and Aging, 17*(2), 299.

Park, D. C., & Reuter-Lorenz, P. (2009). The adaptive brain: Aging and neurocognitive scaffolding. *Annual Review of Psychology, 60*, 173–196.

Patil, A. U., Ghate, S., Madathil, D., Tzeng, O. J. L., Huang, H. W., & Huang, C. M. (2021). Static and dynamic functional connectivity supports the configuration of brain networks associated with creative cognition. *Scientific Reports, 11*, 165.

Pylkkänen, L., Llinás, R., & Murphy, G. L. (2006). The representation of polysemy: MEG evidence. *Journal of Cognitive Neuroscience, 18*(1), 97–109.

Raz, N., Lindenberger, U., Rodrigue, K. M., Kennedy, K. M., Head, D., Williamson, A., Dahle, C., Gerstorf, D., & Acker, J. D. (2005). Regional brain changes in aging healthy adults: General trends, individual differences and modifiers. *Cerebral Cortex, 15*(11), 1676–1689.

Resnick, S. M., Pham, D. L., Kraut, M. A., Zonderman, A. B., & Davatzikos, C. (2003). Longitudinal magnetic resonance imaging studies of older adults: A shrinking brain. *Journal of Neuroscience, 23*(8), 3295–3301.

Robert, C., & Mathey, S. (2007). Aging and lexical inhibition: The effect of orthographic neighborhood frequency in young and older adults. *The Journals of Gerontology: Series B, 62*(6), P340–P342.

Rodd, J., Gaskell, G., & Marslen-Wilson, W. (2002). Making sense of semantic ambiguity: Semantic competition in lexical access. *Journal of Memory and Language, 46*(2), 245–266.

Rodrigue, K. M., & Raz, N. (2004). Shrinkage of the entorhinal cortex over five years predicts memory performance in healthy adults. *Journal of Neuroscience, 24*(4), 956–963.

Rubenstein, H., Garfield, L., & Millikan, J. A. (1970). Homographic entries in the internal lexicon. *Journal of Verbal Learning and Verbal Behavior, 9*(5), 487–494.

Sambataro, F., Murty, V. P., Callicott, J. H., Tan, H. Y., Das, S., Weinberger, D. R., & Mattay, V. S. (2010). Age-related alterations in default mode network: Impact on working memory performance. *Neurobiology of Aging, 31*(5), 839–852.

Steen-Baker, A. A., Ng, S., Payne, B. R., Anderson, C. J., Federmeier, K. D., & Stine-Morrow, E. A. L. (2017). The effects of context on processing words during sentence reading among adults varying in age and literacy skill. *Psychology and Aging, 32*(5), 460–472.

Stern, Y. (2009). Cognitive reserve. *Neuropsychologia, 47*(10), 2015–2028.

Sullivan, E. V., Rohlfing, T., & Pfefferbaum, A. (2010). Quantitative fiber tracking of lateral and interhemispheric white matter systems in normal aging: Relations to timed performance. *Neurobiology of Aging, 31*(3), 464–481.

Tapp, P. D., Siwak, C. T., Gao, F. Q., Chiou, J.-Y., Black, S. E., Head, E., Muggenburg, B. A., Cotman, C. W., Milgram, N. W., & Su, M.-Y. (2004). Frontal lobe volume, function, and β-amyloid pathology in a canine model of aging. *Journal of Neuroscience, 24*(38), 8205–8213.

Vallesi, A., McIntosh, A. R., & Stuss, D. T. (2011). Overrecruitment in the aging brain as a function of task demands: Evidence for a compensatory view. *Journal of Cognitive Neuroscience, 23*(4), 801–815.

Walhovd, K. B., Westlye, L. T., Amlien, I., Espeseth, T., Reinvang, I., Raz, N., Agartz, I., Salat, D. H., Greve, D. N., Fischl, B., Dale, A. M., & Fjell, A. M. (2011). Consistent neuroanatomical age-related volume differences across multiple samples. *Neurobiology of Aging, 32*(5), 916–932.

Wang, L., LaViolette, P., O'Keefe, K., Putcha, D., Bakkour, A., Van Dijk, K. R. A., Pihlajamäki, M., Dickerson, B. C., & Sperling, R. A. (2010). Intrinsic connectivity between the hippocampus and posteromedial cortex predicts memory performance in cognitively intact older individuals. *NeuroImage, 51*(2), 910–917.

Wlotko, E. W., Federmeier, K. D., & Kutas, M. (2012). To predict or not to predict: Age-related differences in the use of sentential context. *Psychology and Aging, 27*(4), 975–988.

Zhou, N., Huang, C. M., Cai, Q., Tzeng, O. J. L., & Huang, H. W. (2022). The effects of aging and perceived loneliness on lexical ambiguity resolution. *Frontiers in Psychology, 13*, Article 978616.

Chapter 11
ERP Evidence for the Modulation of Semantic Transparency on the Recollection of Two-Character Chinese Words

Yi-Jhong Han, Shuo-chieh Huang, and Shih-kuen Cheng

Abstract Event-related potentials (ERPs) were employed to examine whether semantically opaque words were better remembered than were semantically transparent words because of the inconsistency between the meanings of an opaque word and its constituent characters. Experiment 1 employed lexical decisions as the encoding task. The opaque words were better remembered. Moreover, they elicited a greater left parietal ERP old/new effect, an index of recollection-based recognition, than did the transparent words. In Experiments 2 and 3, when semantic transparency judgment was employed as the encoding task to ensure that the constituent characters of the transparent words were processed elaborately to the same degree as those of the opaque words, the recognition advantage for the opaque words was reduced. The inconsistency between a whole word and its constituent characters rendered the opaque words more distinctive and thus better remembered than the transparent words.

Keywords Semantic transparency · Recognition memory · Recollection · ERPs · Left Parietal Old/New Effect

11.1 Introduction

The recognition memory of words is influenced by several linguistic characteristics. One example is that words with a low frequency are more easily recognized compared to those with high frequency (Dobbins et al., 1998; Heathcote et al., 2006). Additionally, concrete words are better remembered compared to abstract words (e.g.,

Y.-J. Han · S. Huang · S. Cheng (✉)
Institute of Cognitive Neuroscience, National Central University, Taoyuan, Taiwan
e-mail: skcheng@cc.ncu.edu.tw

Y.-J. Han
Department of Neuroimaging, Institute of Psychiatry, Psychology and Neuroscience, King's College London, London, UK

D. H. Wu (ed.), *Cognitive and Neural Foundations of Chinese Reading*, Chinese Language Learning Sciences, https://doi.org/10.1007/978-981-96-6669-0_11

Fliessbach et al., 2006; Hamilton & Rajaram, 2001); and nouns are more effectively remembered than verbs (e.g., Gentner, 1981; Kersten & Earles, 2004). Studying how different characteristics of words impact recognition memory has improved our understanding of the distinctiveness effect (Hirshman & Jackson, 1997; Hunt & Elliot, 1980). Glanc and Greene (2007) found that the extent of the recognition memory mirror effect (Glanzer & Adams, 1985, 1990; Joordens & Hockley, 2000) was influenced by the size of a word's orthographic and phonological neighborhood (i.e., words sharing a same rhyme, which is spelled the same for all these words). Their findings supported the premise that words with few neighbors are more distinctive than words with multiple neighbors (Cortese et al., 2004). Similarly, Han et al. (2014) demonstrated that the recognition memory of two-character Chinese words is modulated by semantic transparency, which is defined as the relationship between the meanings of a whole word and its constituents (Libben, 1998; Libben et al., 2003). They also argued that semantically opaque words are more distinctive and thus better remembered than semantically transparent words. In this study, we employed event-related potentials (ERPs) to determine how semantic transparency affects distinctiveness and its interaction with recollection-based recognition.

When a two-character Chinese word's meaning can be inferred from its constituent characters [e.g., "茶(/cha/tea)杯(/bei/cup)," teacup], it is considered semantically transparent. In contrast, semantically opaque words do not have meanings that can be inferred from their constituent characters [e.g., "陽(/yang/sun)春(/chun/spring)," plain]. Han et al. (2014) conducted two experiments that utilized encoding tasks associated with various processing depths. Their results showed that opaque words were more effectively retained in memory than transparent words. The recognition advantage for opaque words was associated with subjective feelings of both recollection and familiarity, which were respectively indexed by the Remember and Know judgments (Rajaram, 1996; Tulving, 1984; Yonelinas & Jacoby, 1995). In the third experiment, Han et al. investigated the effects of semantic transparency and orthographic neighborhood size of two-character words on the Remember judgments of old words. The orthographic neighborhood size was determined by the number of words that share one of the two constituent characters and their positions (Tsai et al., 2006). The results showed that there was an interaction effect between neighborhood size and transparency, whereby the proportion of Remember hits for transparent words was higher for words with fewer neighbors than for words with more neighbors. However, this effect was not observed for opaque words. Han et al. argued that this could be due to varying degrees of competition from neighboring words during retrieval for each type of word. Specifically, transparent words are semantically related to their neighbors as their meanings are all relevant to the constituent characters they share. Thus, a larger neighborhood size could result in more semantically related neighbors competing for retrieval, and consequently, affect the recollection of transparent words. In contrast, the meaning of opaque words is independent from that of their neighbors, resulting in no effect of neighborhood size on their recollection. Han et al. (2014) concluded that the conceptual incongruence between the meanings of a whole word and its constituent characters endowed opaque words with a higher degree of distinctiveness, providing a greater recollection advantage than for transparent words.

The conclusion of Han et al. (2014) should be viewed with some caution due to the following caveats. First, the Remember and Know judgments are self-reports of subjective feelings. Some doubts have been raised about whether Remember judgments reflect a recollection process that is truly independent of high memory strength (see Migo et al., 2012; Wixted & Mickes, 2010 for reviews). An objective index of recollection is required to confirm the contribution of recollection to the recognition advantage for opaque words over transparent words. Second, the absence of a neighborhood size effect on the opaque words is a null result, meaning that it constitutes circumstantial evidence for the role of conceptual incongruence in distinctiveness. To determine whether opaque words gain distinctiveness due to the conceptual incongruence embedded within them, it is necessary to directly manipulate the processing of both the word itself and its constituent characters. Accordingly, we conducted three experiments. Following the approach of Han et al. (2014), we employed lexical decisions as the encoding task in Experiment 1 and recorded ERPs during the test phase. The ERP old/new effects linked to recollection- and familiarity-based recognition were used to examine the mechanism underlying the advantageous recognition of opaque words. Experiment 2 employed semantic transparency judgments as the encoding task to determine whether focusing attention on the relationship between the meanings of a whole word and its constituent characters reduced the recognition advantage of opaque words. Experiment 3 extended the interval between the study and test phases to ensure that the results observed in Experiment 2 were not due to the ceiling effect.

11.2 Experiment 1

This experiment employed mid-frontal and left parietal ERP old/new effects in recognition memory tests to index familiarity- and recollection-based recognition. The mid-frontal effect denotes the positivity associated with hit trials in comparison to correct rejections, with a maximum time range of approximately 300–500 ms over superior frontal sites after stimulus onset. The mid-frontal effect has been linked to familiarity-based recognition because of its relative sensitivity to variables that are assumed to affect familiarity, but not recollection (Azimian-Faridani & Wilding, 2006; Curran, 2000; Woodruff et al., 2006). The left parietal old/new effect denotes the positive-going wave associated with hit trials in comparison to correct rejections. This effect peaks over the left temporoparietal region and lasts approximately 400–600 ms after stimulus onset. The left parietal old/new effect has been linked to recollection-based recognition because it is sensitive to variables that are assumed to affect recollection, but not familiarity (Curran, 2000; Düzel et al., 1997; Rugg & Curran, 2007; Wilding & Rugg, 1996). If both recollection and familiarity contribute to the advantageous recognition of opaque words, these words should yield greater left parietal and mid-frontal effects than do transparent words.

11.2.1 Materials and Methods

11.2.1.1 Participants

Twenty-nine students (12 men; aged between 18 and 24 years) attending National Central University in Taoyuan, Taiwan, were recruited. All the participants were right-handed, native Mandarin Chinese speakers with normal or corrected-to-normal vision. The data of nine participants were disregarded because they contributed an insufficient number of valid ERP trials (<16) in at least one critical experimental condition. Each participant received NT$500 as compensation. Written informed consent was obtained from all the participants prior to study commencement.

11.2.1.2 Materials

This experiment utilized stimuli from Han et al. (2014), which consisted of 184 two-character Chinese words selected from the Academia Sinica Balanced Corpus (Chen et al., 1996, p. 199; Huang et al., 1998). Half of the words were transparent, and the other half were opaque. A 5-point transparency rating on these words was obtained from 24 university students who were not involved in the current study. Results indicated that transparent words had significantly higher transparency ratings (mean = 4.32, SD = 0.36) than opaque words (mean = 1.55, SD = 0.45; t(182) = 46,35, $p < 0.001$). The frequencies of transparent and opaque words were statistically equivalent, and neighborhood sizes for both types of words were also matched. Additionally, a 7-point concreteness rating was obtained from 20 university students who did not participate in this study. Results indicated that there was no significant difference in concreteness between transparent and opaque words. The lexical decision task included 92 two-character nonwords, which were generated by rearranging real two-character words that were not used in the experiment. The nonwords were not included in the recognition memory test.

11.2.1.3 Procedure

The experimental procedure was similar to that of Han et al. (2014), with the exception that ERPs were recorded and the Remember/Know procedure (Rajaram, 1996; Tulving, 1984; Yonelinas & Jacoby, 1995) was not employed during the test phase. Participants were fitted with an elastic electrode cap and seated in front of a computer monitor at a viewing distance of 90 cm. In the study phase, participants made lexical decisions to 46 transparent words, 46 opaque words, and 92 nonwords. Each study trial commenced with a fixation cross that was presented in the center of the screen for 500 ms, followed by a two-character item displayed for 300 ms. A second fixation character "@" was then presented for 1700 ms, and during this 2000-ms period, participants were instructed to determine whether the item was a real word or a nonword. A blank screen was presented for 500 ms before the start of the next trial. The order of presentation of the transparent words, opaque words, and nonwords was randomized for each participant, and the participants provided responses by pressing one of two response keys with the index finger of each hand, with the mapping of the hands to the response categories counterbalanced across participants. During the study phase, the participants were not instructed to memorize the presented items, nor were they informed of the subsequent recognition test. Four filler trials were inserted at both the beginning and the end of the study phase.

Following the study phase, participants engaged in a 5-min video game distraction task, after which they underwent a test phase where they made old/new judgments to the 92 real words that had been presented in the lexical decision task and 92 new real words. Each test trial started with a fixation cross presented at the center of the screen for 500 ms, followed by the test item displayed for 300 ms. A second fixation character "@" was then presented for 2200 ms, during which the participants made old/new judgments with their index fingers. The mapping of the hands to response categories (old vs. new) was counterbalanced across the participants. The new words consisted of 46 transparent words and 46 opaque words, and the designation of the real words as old versus new words was counterbalanced across the participants.

11.2.1.4 EEG Recording and ERP Analysis

Electroencephalography (EEG) was recorded during both the study and test phases, but only those recorded in the test phase are reported in this paper. The recordings were made using 64 silver/silver chloride electrodes, 62 of which were embedded in an elastic cap (Quick-Cap, Neuromedical Supplies, Sterling, Texas, USA). The remaining two electrodes were placed on the right and left mastoids. All channels were referenced to a channel located between the Cz and CPz electrode sites. Vertical and horizontal electro-oculograms were recorded bipolarly from electrodes placed superior and inferior to the right eye and on the outer canthi of each eye. Data were sampled at 250 Hz and digitized with 32-bit resolution. All channels were amplified using SYNAMPS2 (Neuroscan, Inc., El Paso, Texas, USA), with a band-pass of 0.05–70 Hz (3 dB points). Interelectrode impedance was maintained below 5 kΩ.

The EEG data were analyzed using the open-source toolboxes of EEGLAB (Delorme & Makeig, 2004) and ERPLAB (Lopez-Calderon & Luck, 2014). A low-pass filter with a cut-off frequency of 30 Hz (half-amplitude point, 48 dB/octave) was first applied to the EEG data. The data were then re-referenced to the average of the EEG data obtained for the two mastoids and segmented into 2044-ms epochs, starting from 100 ms before the onset of the test items. All the epochs were then baseline-corrected with the pre-stimulus interval and subjected to independent component analysis to identify and remove the components of vertical and horizontal eye movements. Epochs containing signals exceeding 90 μV from the baseline in any channel were rejected using a simple artifact rejection threshold supplied by ERPLAB.

11.2.2 Behavioral Results

Study Phase. The mean (SD) values of the accuracy rates of the lexical decision judgments regarding the transparent words, opaque words, and nonwords were 0.98 (0.03), 0.96 (0.03), and 0.97 (0.05), respectively. One-way repeated-measures analysis of variance (ANOVA) revealed that the accuracy rates of the three stimulus types did not differ significantly, $F(1.54, 29.28) = 1.235$, $p = 0.296$, $\eta_p^2 = 0.061$. ANOVA of the response times of accurate responses to the transparent words ($M = 648$ ms, $SD = 105$ ms), opaque words ($M = 659$ ms, $SD = 100$ ms), and nonwords ($M = 738$ ms, $SD = 116$ ms) indicated that the effect of the stimulus type was significant, $F(1.56, 29.59) = 52.071$, $p < 0.001$, $\eta_p^2 = 0.733$. Post hoc analyses with Bonferroni correction demonstrated that the response times for the transparent and opaque words were significantly shorter than the response time for the nonwords (both p values < 0.001). However, the response times for the transparent and opaque words did not differ significantly ($p = 0.286$).

Test Phase. Table 11.1 displays the proportion of old and new responses to the two types of test words. Repeated-measures ANOVA including the test item type (old vs. new) and transparency (transparent vs. opaque) as factors to the old response rates revealed that the main effects of the test item type and transparency were significant, $F(1, 19) = 548.926$, $p < 0.001$, $\eta_p^2 = 0.967$ and $F(1, 19) = 18.235$, $p < 0.001$, $\eta_p^2 = 0.490$, respectively. This result revealed that the hit rate was higher than the false alarm rate, and that the opaque words received more old responses than did the transparent words. Notably, the interaction between the test item type and transparency was also significant, $F(1, 19) = 31.41$, $p < 0.001$, $\eta_p^2 = 0.623$, suggesting that the difference between the hit rates for the opaque and transparent words was higher than that between the corresponding false alarm rates. A follow-up simple main effect analysis indicated that the hit rate for the opaque words was significantly higher than that for the transparent words, $F(1, 19) = 33.413$, $p < 0.001$, $\eta_p^2 = 0.637$. By contrast, the difference between the false alarm rates for these two types of words was not significant, $F(1, 19) = 3.481$, $p = 0.078$, $\eta_p^2 = 0.155$.

11.2.3 ERP Results

Averaged ERPs were calculated for the hits and correct rejections to the opaque and transparent words (hereafter referred to as opaque hits, opaque correct rejections, transparent hits, and transparent correct rejections, respectively; Fig. 11.1). The mean (range) values of the number of trials for the opaque hits, opaque correct rejections, transparent hits, and transparent correct rejections in the ERP analysis were 36 (20–44), 37 (17–45), 37 (28–45), and 40 (16–50), respectively. ERPs were quantified by measuring the mean amplitudes of time intervals of 300–500 and 500–800 ms. These

Table 11.1 Mean proportions (with SDs in parentheses) of the Old responses (Experiments 1 to 3) and Remember/Know judgments (Experiment 3) to the opaque and transparent test words as a function of old/new status

	Old items (Hit)		New items (False alarm)	
	Transparent	Opaque	Transparent	Opaque
Experiment 1				
Old response	0.74 (0.11)	0.86 (0.12)	0.13 (0.09)	0.11 (0.10)
Experiment 2				
Old response	0.91 (0.08)	0.91 (0.05)	0.10 (0.15)	0.08 (0.13)
Experiment 3				
Old response	0.73 (0.14)	0.75 (0.18)	0.14 (0.15)	0.14 (0.15)
Remember	0.54 (0.19)	0.52 (0.09)	0.08 (0.09)	0.06 (0.09)
Know	0.19 (0.15)	0.23 (0.14)	0.07 (0.08)	0.08 (0.08)

intervals approximately accord with the intervals used in other studies (Azimian-Faridani & Wilding, 2006; Paller et al., 2007; Woodruff et al., 2006) examining ERP old/new effects corresponding to the mid-frontal and left parietal regions.

For the analysis of the mid-frontal effect, repeated-measures ANOVA was conducted on the data from the midanterior (F1, Fz, F2), midcentral (C1, Cz, C2), and midposterior (P1, Pz, P2) recording sites with transparency (transparent and opaque), response category (hit and CR), and caudality (anterior, central, and posterior) as factors. Greenhouse–Geisser correction was applied, where applicable, despite the degrees of freedom reported being uncorrected. The main effect of the response category was significant, $F(1, 19) = 20.368, p < 0.001, \eta_p^2 = 0.517$. The interactions of the response category and transparency, $F(1, 19) = 1.990, p = 0.175, \eta_p^2 = 0.095$, or of transparency, response category, and causality $F(2, 38) = 0.018, p = 0.920, \eta_p^2 < 0.001$. The findings suggested that the ERPs associated with hits were more positive than those associated with correct rejections. No evidence indicated that the mid-frontal old/new effect corresponding to the transparent and opaque words differed significantly.

The analysis of the left parietal old/new effect was conducted on the data from the left frontal (F3, F5, F7, FC3, FC5, FT7), right frontal (F4, F6, F8, FC4, FC6, and FT8), left parietal (P3, P5, P7, CP3, CP5, TP7), and right parietal (P4, P6, P7, CP4, CP6, TP7) recording sties. The factors considered were transparency, response category, hemisphere (left and right), and caudality (anterior and posterior). The main effect of the response category was significant, $F(1, 19) = 6.227, p = 0.022, \eta_p^2 = 0.247$ (Fig. 11.2), reflecting the more positive-going waveforms associated with the hit trials as opposed to the correct rejection trials. The interaction between the response category and transparency was also significant, $F(1, 19) = 4.593, p = 0.045, \eta_p^2 = 0.195$. Notably, the four-way interaction among transparency, response category, caudality, and hemisphere was significant, $F(1, 19) = 6.577, p = 0.019, \eta_p^2 = 0.257$, suggesting a difference between the old/new effects of the two types of words that varied across the scalp regions. Follow-up analyses by caudality revealed that the

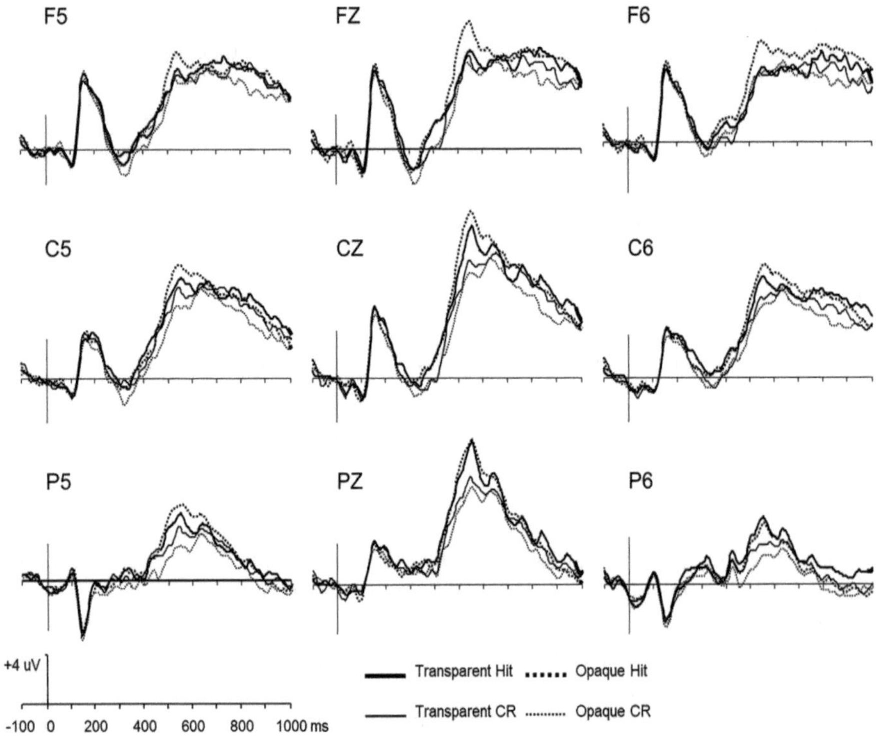

Fig. 11.1 ERP waveforms associated with the hits and correct rejections to the transparent and opaque words in Experiment 1

three-way interaction among transparency, response category, and hemisphere was significant over the posterior scalp region, $F(1, 19) = 6.929, p = 0.016, \eta_p^2 = 0.267$, but was nonsignificant over the anterior scalp region, $F(1, 19) = 0.204, p = 0.657, \eta_p^2 = 0.011$. Further analyses by hemisphere of the posterior region revealed that the transparency–response category interaction was significant over the left posterior scalp region, $F(1, 19) = 8.044, p = 0.011, \eta_p^2 = 0.297$, but was nonsignificant over the right posterior scalp region, $F(1, 19) = 1.046, p = 0.319, \eta_p^2 = 0.052$. However, the main effect of the response category was significant over the right posterior region, $F(1, 19) = 5.935, p = 0.025, \eta_p^2 = 0.238$. The result demonstrated that, regardless of transparency, hits elicited more positive-going amplitudes than did correct rejections in this region. These findings suggested that the left magnitude of the parietal old/new effect was greater for the opaque words than for the transparent words. The left parietal old/new effect (between the hits and correct rejections) was significant for the opaque words, $F(1, 19) = 27.649, p < 0.001, \eta_p^2 = 0.593$, but was nonsignificant for the transparent words, $F(1, 19) = 0.935, p = 0.346, \eta_p^2 = 0.047$. Another perspective on the two-way interaction was that the difference between the correct rejections for the opaque and transparent words was greater than the difference between the hits for the two types of words. The waveforms associated with the correct rejections were

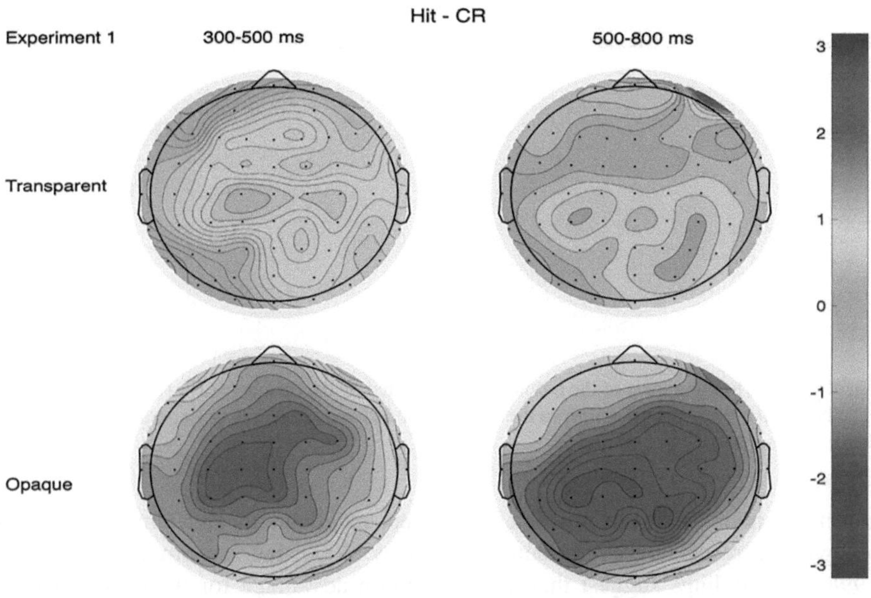

Fig. 11.2 The Topographies of the old/new effect elicited by transparent (upper panel) and opaque (lower panel) words in Experiment 1, derived by subtracting the waveforms of correct rejections from the hits, in the time windows of 300–500 ms (left) and 500–800 ms (right)

marginally more positive going for the transparent words than for the opaque words, F(1, 19) = 4.365, $p = 0.050$, $\eta_p^2 = 0.187$. By contrast, the waveforms associated with the hits for these two types of words were not statistically different, F(1, 19) = 1.315, $p = 0.266$, $\eta_p^2 = 0.065$. Further topography analysis (normalized using the max–min method; McCarthy & Wood, 1985) suggested no evidence of differences in topographies for opaque hits versus correct rejections and for transparent versus opaque correct rejections, F(1, 65) = 1.128, $p = 0.230$, $\eta_p^2 = 0.056$ (Fig. 11.3).

11.2.4 Discussion

The behavioral results in this study partially replicate the findings of Han et al. (2014), demonstrating that opaque words are better recognized than transparent words. The hit rate was significantly higher for the opaque words than for the transparent words. Unlike the mirror effect (Glanzer & Adams, 1985) reported in Experiment 1 of the study by Han et al. (2014). The false alarm rates were not statistically different for the opaque and transparent words in the current experiment. Notably, the false alarm component of the mirror effect was also nonsignificant in Experiments 2 and 3 of that study, despite more false alarms being provided for the transparent words than for the opaque words in both studies. Following the argument that the hit and false

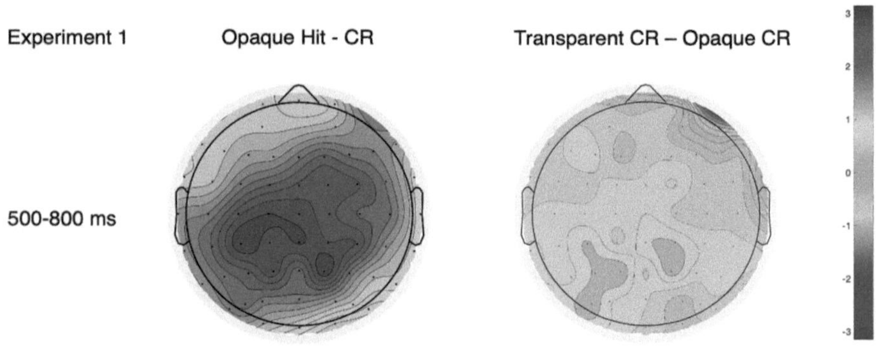

Fig. 11.3 The topographies of the old/new effect elicited by the opaque words (left) and the difference between the correct rejections to the transparent and opaque words in Experiment 1 (right) in the time window of 500–800 ms.

alarm components of the mirror effect reflect recollection- and familiarity-based recognition, respectively (Joordens & Hockley, 2000; but see Stretch & Wixted, 1998), our findings suggest that the recognition advantage for the opaque words mainly results from the recollection-based recognition of old items. The difference in recognition performance for the opaque and transparent words is unlikely due to varying encoding levels because the accuracy and speed of the lexical decision judgments for these two types of words were statistically comparable. The ERP results support that the recognition advantage for the opaque words is mainly based on recollection-based recognition. The magnitude of the mid-frontal ERP old/new effect for the opaque and transparent words was comparable, reflecting a similar contribution of familiarity to the recognition of these two types of words. By contrast, the magnitude of the left parietal old/new effect was greater for the opaque words than for the transparent words, reflecting a greater contribution of recollection to the identification of the opaque words. In sum, the ERP findings provide further evidence for the contribution of recollection to the recognition advantage for opaque words.

11.3 Experiment 2

The results of Experiment 1 support the argument that recollection is essential for the recognition advantage for opaque words. However, it remains unclear whether it was the conceptual relationship between the meanings of a two-character word and its constituent characters that resulted in the discrepant recognition performance for the opaque and transparent words. Experiment 2 was conducted to answer this question, employing semantic transparency judgments as the encoding task. Accomplishing semantic transparency judgments requires explicitly processing the meanings of a whole word and its constituent characters. With comparable levels of attentional resources being allocated to the transparent words and opaque words, the recognition advantage for the opaque words could be reduced. As in Experiment 1, the mid-frontal and left parietal ERP old/new effects were used as indices of familiarity- and recollection-based recognition, respectively, for the opaque and transparent words. We predicted that under such direct manipulation, the hit rates and the mid-frontal effect and the left parietal effect would be statistically comparable for the two types of words.

11.3.1 Materials and Methods

11.3.1.1 Participants

Twenty-two National Central University students (12 men; aged between 18 and 26 years) who did not participate in Experiment 1 were recruited. All the participants were right-handed, native Mandarin Chinese speakers with normal or corrected-to-normal vision. The data of two participants were disregarded because they contributed an insufficient number of valid ERP trials (<16) in at least one critical experimental condition. Each participant received NT$500 as compensation. Written informed consent was obtained from all participants prior to study commencement.

11.3.1.2 Materials

The transparent and opaque words used in Experiment 1 were employed as stimuli in Experiment 2. As in Experiment 1, half of the words were shown in the study phase and were then presented as old items in the subsequent test phase. The other half of the words served as new items in the test phase. For both the old and new words, half of the words were opaque and transparent, respectively. The designation of the old and new words was counterbalanced across the participants.

11.3.1.3 Procedure

The procedure was similar to that of Experiment 1, with the exception that a semantic transparency judgment was employed as the encoding task during the study phase. The participants were instructed to judge, on the presentation of each study word, whether the meaning of the whole word could be derived from the meanings of the two constituent characters. The presentation order of the transparent words and opaque words was randomized for each participant. The participants provided responses by pressing one of two response keys with the index finger of each hand. The mapping of the hands to the response categories (transparent vs. opaque) was counterbalanced across the participants. During the study phase, the participants were not instructed to memorize the presented items. Four filler trials were inserted at both the beginning and end of the study phase. The ERP recording procedure and data processing procedures were identical to those used in Experiment 1.

11.3.2 Behavioral Results

Study phase. The accuracy rates of the transparency judgments of the transparent words (M = 0.819, SD = 0.152) and the opaque words (M = 0.879, SD = 0.074) did not differ significantly, $t(19) = 1.710, p = 0.104$. Two-way ANOVA with the stimulus type (transparent vs. opaque) and response accuracy (accurate vs. inaccurate) as factors revealed that accurate responses were entered significantly more quickly than were inaccurate responses, $F(1, 19) = 26.500, p < 0.001, \eta_p^2 = 0.582$; 1089 vs. 1231 ms. The main effect of the stimulus type was nonsignificant, $F(1, 19) = 0.534$, $p = 0.474$, $\eta_p^2 = 0.027$, nor was its interaction with response accuracy, $F(1, 19) = 0.118$, $p = 0.735$, $\eta_p^2 = 0.006$.

Test Phase. Table 11.1 displays the proportion of old and new responses for the two types of test words. Repeated-measures ANOVA with the test item type (old vs. new) and transparency (transparent vs. opaque) for the old response rates as factors revealed that the main effect of the test item type was significant, $F(1, 19) = 538.842$, $p < 0.001, \eta_p^2 = 0.966$, indicating that the hit rate was higher than the false alarm rate. The main effect of transparency was nonsignificant, $F(1, 19) = 0.614$, $p = 0.443$, $\eta_p^2 = 0.031$, nor was its interaction with test item type, $F(1, 19) = 1.276, p = 0.273$, $\eta_p^2 = 0.063$.

11.3.3 ERP Results

ERPs were quantified by measuring the mean amplitudes of the time intervals of 300–500 and 500–800 ms corresponding to the mid-frontal and left parietal ERP old/new effects, respectively. The mean (range) values of the number of trials corresponding to the opaque hits, opaque correct rejections, transparent hits, and transparent correct rejections were 36 (20–44), 37 (17–45), 37 (28–45), and 36 (16–45), respectively.

The analysis of the mid-frontal effect in Experiments 1 and 2 was conducted using the same approach. The mean amplitudes from frontal (F1, Fz, F2), central (C1, Cz, C2), and posterior (P1, Pz, P2) electrodes were used for the analysis. Electrode sites within each causality were averaged. The main effect of the response category was significant, $F(1, 19) = 20.608$, $p < 0.001$, $\eta_p^2 = 0.520$, but its interaction with caudality was nonsignificant, $F(2, 38) = 0.835$, $p = 0.402$, $\eta_p^2 = 0.042$ (Fig. 11.4). This result suggested that the waveforms associated with the hits were more positive going than were correct rejections across the three scalp regions (see Fig. 11.5). The main effect of transparency was nonsignificant, $F(1, 19) = 0.548$, $p = 0.468$, $\eta_p^2 = 0.028$, nor was its interaction with the response category, $F(1, 19) = 0.425$, p

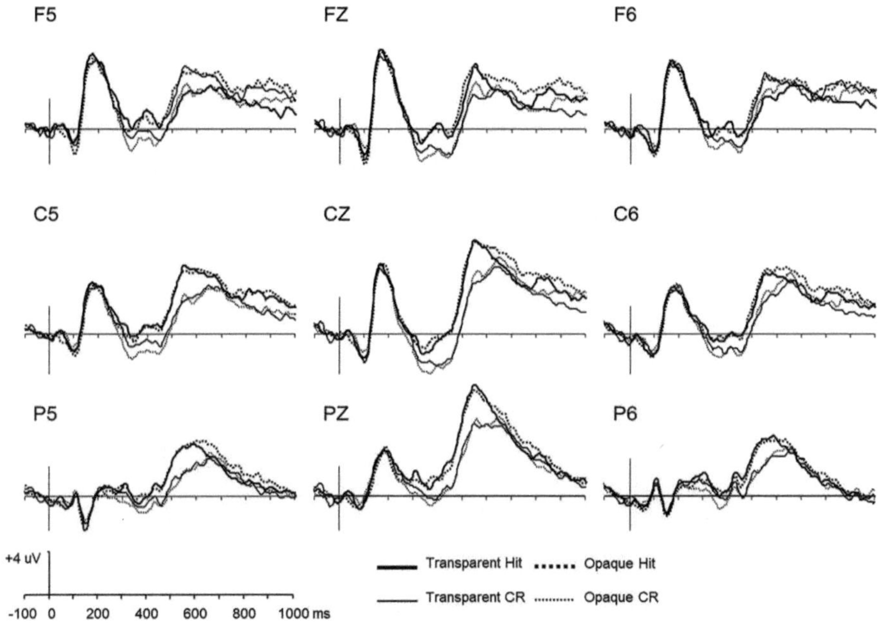

Fig. 11.4 ERP waveforms associated with the hits and correct rejections to the transparent and opaque words in Experiment 2

$= 0.522$, $\eta_p^2 = 0.022$. The interaction among transparency, response category, and caudality was also nonsignificant, $F(2, 38) = 0.251$, $p = 0.702$, $\eta_p^2 = 0.013$. The results indicate that the ERP old/new effect is observable for both the transparent and opaque words in the 300–500-ms time window.

As in Experiment 1, the analysis of the left parietal old/new effect was conducted on the data from the left frontal (F3, F5, F7, FC3, FC5, FT7), right frontal (F4, F6, F8, FC4, FC6, FT8), left parietal (P3, P5, P7, CP3, CP5, TP7), and right parietal (P4, P6, P7, CP4, CP6, TP7) recording sties with the factors of transparency, response category, hemisphere (left vs. right), and caudality (anterior vs. posterior). Electrode sites within each causality were averaged. The mean main effect of the response category was significant, $F(1, 19) = 6.077$, $p = 0.023$, $\eta_p^2 = 0.242$. No other significant main effects of interaction were found (all F values < 2.2). The results indicate that the ERP old/new effect is observable for all words (whether transparent or opaque) in the 500–800-ms time window. In sum, different from what was observed in Experiment 1, the results of Experiment 2 did not indicate a greater parietal old/new effect for the opaque words than for the transparent words.

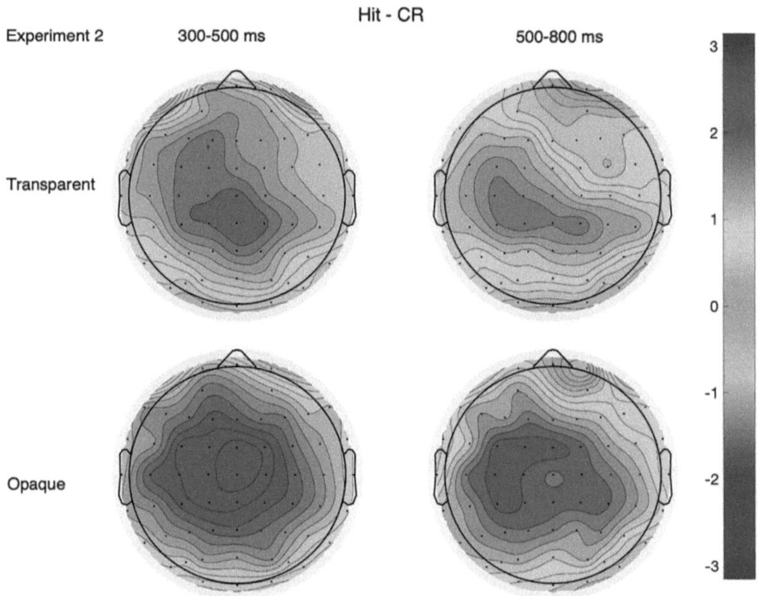

Fig. 11.5 The Topographies of the old/new effect elicited by transparent (upper panel) and opaque (lower panel) words in Experiment 1, derived by subtracting the waveforms of correct rejections from the hits, in the time windows of 300–500 ms (left) and 500–800 ms (right)

11.3.4 Discussion

As predicted, with transparency judgement as the encoding task, the participants' transparency judgments of the transparent and opaque words were equally accurate in terms of both accuracy and response time. Notably, compared with when lexical decisions were used as the encoding task, the transparent words were remembered better and to the same degree as were the opaque words. Moreover, both transparent and opaque words elicited significant mid-frontal and left parietal old/new effects. However, the ERP old/new effects did not differ between the transparent and opaque words.

The findings provide further evidence to support the argument of Han et al. (2014). They reported that with transparency judgment as the encoding task, transparent and opaque words induced comparable hit rates and false alarm rates. Notably, the mid-frontal and left parietal effects elicited by both types of words were also comparable, indicating the equivalent contributions of recollection and familiarity. Compared with the findings in Experiment 1 (with a lexical decision encoding task), the results of Experiment 2 suggested that focusing on the relationship between a word and its constituents minimized the difference in the recollection of the transparent and opaque words, as well as abolished the distinctiveness of opaque words.

11.4 Experiment 3

In Experiment 2, memory performance corresponding to the transparent and opaque words was statistically equivalent. However, the high hit rates (>90%) raised the concern that this equivalence was due to the ceiling effect rather than due to the manipulation associated with the encoding task. We conducted Experiment 3 to further test memory performance under the same encoding condition as that in Experiment 2, but to reduce overall memory performance to exclude the possibility of a ceiling effect. This can be achieved by manipulating divided attention (Gardiner & Parkin, 1990), the exposure time during encoding (Yonelinas, 1994), the number of repetitions (Stretch & Wixted, 1998), or study–test delay (Gardiner & Java, 1991). The study–test delay offers a straightforward manipulation for the current study to maintain the design, with the benefit of reducing overall memory performance. The delay between the study and test phases was increased from 5 min, as in Experiment 2, to 24 h in Experiment 3. Recollection tends to decay more sharply than familiarity over time (Gardiner & Java, 1991). Nevertheless, extending the study–test delay is expected to reduce the memory performance of both transparent and opaque words. Moreover, the Remember/Know procedure was employed during the test phase to determine whether the contribution of recollection to the opaque words and transparent words was comparable (as suggested by the equivalent left parietal ERP old/new effects to the two types of words in Experiment 2).

11.4.1 Materials and Methods

11.4.1.1 Participants

Sixteen National Central University students (four men; aged between 18 and 25 years) who did not participate in the first two experiments were recruited. All the participants were right-handed, native Mandarin Chinese speakers with normal or corrected-to-normal vision. Each participant received NT$200 as compensation. Written informed consent was obtained from all participants prior to study commencement.

11.4.1.2 Materials

The stimuli were the transparent and opaque words used in Experiment 1 and Experiment 2. As in those experiments, half of the words were shown in the study phase and then presented as old items in the subsequent test phase. The other half of the words served as new items in the test phase. For both the old and new words, half of the words were opaque and transparent, respectively. The designation of the words as old and new words was counterbalanced across the participants.

11.4.1.3 Procedure

The procedure was the same as that in Experiment 2, except that Experiment 3 featured no EEG recordings and involved a study–test interval of 24 h rather than 5 min. The mean (range) of the study–test delay in the participants was 24.09 (19–30.5) hours. Following the approach of Han et al. (2014), the Remember/Know procedure (Rajaram, 1996; Tulving, 1984; Yonelinas & Jacoby, 1995) was employed to examine the subjective experiences associated with the mnemonic judgments.

11.4.2 Results and Discussion

Study phase. The accuracy rates of the transparency judgments of the transparent words (M = 0.826, SD = 0.179) and the opaque words (M = 0.896, SD = 0.048) did not differ significantly, $t(15) = 1.490, p = 0.157$. Two-way ANOVA with the stimulus type (transparent vs. opaque) and response accuracy (accurate vs. inaccurate) as factors revealed no significant main effects of the stimulus type, $F(1, 15) = 0.718$, $p = 0.410$, $\eta_p^2 = 0.046$ (1128 vs. 1152 ms) or response accuracy, $F(1, 15) = 0.440$, $p = 0.517$, $\eta_p^2 = 0.028$ (1149 vs. 1130 ms). However, a significant interaction was found between the two factors, $F(1, 15) = 14.691$, $p = 0.002$, $\eta_p^2 = 0.495$. A simple main effect analysis indicated that accurate responses corresponding to the transparent words were made significantly more quickly than those corresponding to the opaque words, $F(1, 15) = 11.786$, $p = 0.004$, $\eta_p^2 = 0.440$ (1074 vs. 1229 ms). By contrast, inaccurate responses corresponding to the transparent words were made significantly more slowly than were those corresponding to the opaque words, $F(1, 15) = 6.73$, $p = 0.020$, $\eta_p^2 = 0.310$ (1187 ms vs. 1074 ms).

Test Phase. Table 11.1 displays the proportion of old and Remember/Know responses to the two types of words. Three repeated-measures ANOVAs with the test item type (old vs. new) and transparency (transparent vs. opaque) for the old, Remember and Know response rates as factors were conducted. The analysis of the old responses indicated that the main effect of the test item type was significant, $F(1, 15) = 226.313$, $p < 0.001$, $\eta_p^2 = 0.938$, revealing that the hit rate was higher than the false alarm rate. The main effect of transparency was nonsignificant, $F(1, 15) = 0.145$, $p = 0.709$, $\eta_p^2 = 0.010$, nor was its interaction with the test item type, $F(1, 15) = 0.276$, $p = 0.607$, $\eta_p^2 = 0.018$.

A similar pattern was found in the analysis of Remember responses—the main effect of the test item type was significant, $F(1, 15) = 156.116$, $p < 0.001$, $\eta_p^2 = 0.912$, indicating that the proportion of Remember responses was higher for old items than for new items. The main effect of transparency was nonsignificant, $F(1, 15) = 1.541$, $p = 0.234$, $\eta_p^2 = 0.093$, as was the interaction between transparency and the test item type, $F(1, 15) = 0.007$, $p = 0.933$, $\eta_p^2 < 0.001$. The Know responses were not analyzed further due to the analysis was focused on recollection.

The 24-h delay between the study and test phases in Experiment 3 resulted in poorer old/new recognition performance compared with that in Experiment

2. Notably, the recognition performance, in terms of old/new recognition or recollection-based recognition, indexed by the proportion of Remember responses, remained statistically identical for the opaque and transparent words. Therefore, the equivalent memory performance for the two types of words observed in Experiment 2 was not due to the ceiling effect.

11.5 General Discussion

To recapitulate, the findings of Experiment 1 are consistent with the findings of Han et al. (2014), demonstrating that opaque words are better remembered than are transparent words. In this study, a greater left parietal ERP old/new effect was observed for the opaque words. Moreover, the mid-frontal old/new effect for the two types of words was equivalent. This result indicated that the memory advantage for the opaque words was associated with recollection, but not with familiarity. This premise is consistent with our previous finding that opaque words were remembered to a greater extent than were transparent words, as indexed by the Remember/Know procedure in Han et al. (2014). Experiment 2 and Experiment 3 further indicated that the recognition advantage for the opaque words was reduced when the transparency judgment was employed as the encoding task. The observation suggests that the processing of the meanings of the whole word and its constituents contributes crucially to the recognition advantage for opaque words.

It is unclear why a recognition advantage existed for opaque words over transparent words when the encoding task was a lexical decision, but not for a transparency judgment. Multiple psycholinguistic and neurolinguistics studies have highlighted the various morphological processes involved in compound words. Researchers have demonstrated that when processing opaque words, both the meanings of the whole words and their constituents are activated (e.g., Lee et al., 2021; MacGregor & Shtyrov, 2013), although the activation of the whole word might be stronger than the activation of the constituents. As asserted by Han et al. (2014), the inconsistency between the meanings of a whole word and its constituents renders opaque words salient. The probable mechanism is that attention is drawn to the competition between the inconsistency in the activation of the whole words and the constituents. This could explain why the opaque words were better remembered than the transparent words when the encoding task was a lexical decision task. Han et al. (2014) indirectly addressed this question by examining the interaction between transparency and neighborhood size, and they argued that the inconsistency between the meanings of the whole word and the constituent characters rendered an opaque word distinctive and was thus better remembered than transparent words. Notably, the meanings of the whole word and the constituents in the transparent words were congruent, but not identical. When the transparency judgment task was employed as the encoding task, the participants were required to scrutinize the subtle difference between the meaning of the whole word and its constituents to the same extent for the transparent and opaque words. This offset the recognition advantage for the opaque words. The presence and absence of the recognition advantage for the opaque words, respectively observed when lexical decision and transparency judgment were employed as the encoding tasks, provide unequivocal evidence that the recognition advantage for opaque words results from the conceptual inconsistency between the whole word and the constituents.

One novel finding is that the greater left parietal old/new effect for the opaque words observed in Experiment 1 was driven by the differences between the correct rejections, but not the hits for the two types of words. Specifically, the waveforms associated with the correct rejections to the transparent words were more positive-going than were the correct rejections to the opaque words over the left parietal scalp region. However, no significant difference was detected between the waveforms associated with the hits for the opaque and transparent words. The left parietal old/new effect has been employed to index recollection in multiple studies. Most of these studies have focused on the waveforms associated with the hits in different conditions (e.g., correct source vs. incorrect source; Rugg et al., 1998b) or deep encoding versus shallow encoding (Rugg et al., 1998a). The new items in those studies were not differentiated into distinct types; thus, no study has examined the contribution of correct rejections to the parietal effect. Therefore, our finding of transparent and opaque correct rejections modulating the parietal effect merits further discussion. Two interpretations of this finding are presented as follows.

First, it could be argued that the positivity associated with the correct rejections for the transparent words as opposed to the opaque words resembles the left parietal old/new effect conventionally reported in the comparison between hits and correct rejections. As shown in Fig. 11.3, the temporal duration of this positive effect resembles the hit-versus-correct rejection parietal effect observed in other studies. Moreover, a topographic analysis found no difference in the distribution of the correct rejections between the transparent and opaque words and in of the left parietal effect of the opaque words. If the positivity associated with transparent correct rejections as opposed to opaque correct rejections was the recollection-based left parietal old/new effect, what further information was recollected in response to the transparent new words compared with in response to the opaque words, and why? One possibility is that, in comparison to opaque words, transparent words are more likely to induce recall-to-reject processing (Rotello & Heit, 1999). Several studies have reported on the elicitation of the parietal effect by the recall-to-reject processing of correct rejections. For example, the parietal old/new effect has been observed in correct rejections for the nontargets of an exclusion task (e.g., Herron & Rugg, 2003) and in correct rejections for the similar lures of the Deese–Roediger–McDermott (DRM) task (Morcom, 2015). The positivity associated with the transparent new words might therefore reflect the recall and rejection of a cohort of words that share semantic features with the test items as compared with the opaque words.

Despite its appeal, the recall-to-rejection interpretation is not a compatible explanation for the results of the current investigation. Specifically, the new words employed, whether opaque or transparent, were all semantically unrelated to the old words. Specifically, none of the two-character words used shared characters with the old words. Therefore, a low likelihood exists that more old transparent words than old opaque words were recalled in the correct rejection trials to produce the parietal effect. Another challenge to employing the recall-to-reject interpretation is why the differences between the correct rejections for the two types of words disappeared in Experiment 2, where the transparency judgment task was employed as the encoding task. How the recollection differences due to the linguistic characteristics of new test items were modulated by the encoding task is unclear.

The second interpretation is that the difference between the waveforms associated with opaque and transparent correct rejections reflected the processing of retrieval orientation. Successful retrieval relies not only on the presence of retrieval cues but also on appropriate interactions between cues and memory representations. Both the principles of encoding specificity (Tulving & Thomson, 1973) and transfer-appropriate processing (Morris et al., 1977) emphasize that memory performance depends on the degree to which the context or cognitive operations engaged during encoding are recapitulated during retrieval. An implication of these principles is that people can vary the manner in which a retrieval cue is processed to optimize its compatibility with a target memory representation. The objective is to enhance the probability of retrieval success (Hornberger et al., 2004; Rugg & Wilding, 2000). The neural correlates of retrieval orientation can be revealed by comparing the brain activities elicited by the same retrieval cues used to probe different types of memories (Rugg & Wilding, 2000). The correctly rejected new items in a recognition test are considered the optimal retrieval cues for examining the retrieval orientation effect because the influence of the successful retrieval of different types of memory would be minimal. Correct rejection trials have been found to be associated with different patterns of ERPs when the new items were accompanied by old items that were of a different nature or from different encoding contexts. For instance, in a previous study, the waveforms associated with the correct rejection trials were more positive going over the anterior central scalp region when the old items were shallowly encoded as opposed to when the old items were deeply encoded in the study phase (Rugg et al., 2000). Researchers have further demonstrated that the processing of retrieval cues can be affected by the input modality (e.g., visual vs. auditory) of the sought-for memory. In this context, alternative interpretations of retrieval efforts and study–test cue overlapping have been disregarded (Hornberger et al., 2004; Robb & Rugg, 2002) in favor of the retrieval orientation effect. Therefore, the positivity associated with the transparent as opposed to opaque correct rejection trials reflects the distinct forms of retrieval orientation processing induced by the linguistic nature of opaque and transparent words. The retrieval orientation interpretation is advantageous over the recall-to-reject interpretation in that the former interpretation accommodates the question of why the positivity associated with opaque correct rejections was not observed in Experiment 2. Multiple studies have demonstrated that the processing of retrieval cues can be modulated by the levels of processing or the modality under which the sought-for information is encoded during the study phase. The presence and absence of the transparent–opaque correct rejection differences in Experiment 1 and Experiment 2 might demonstrate retrieval orientation processing induced by the distinct encoding processes of the transparent versus opaque words. Specifically, when the words were encoded with lexical decisions, the transparent words may have elicited the recollection of contextual information to a greater extent. The rationale is that transparent words possess conceptual fluency attributable to the similar representations of the whole word and its constituents (e.g., "teacup" shares the semantic concepts of "tea" and "cup"). However, this conceptual fluency was not observed when a semantically oriented encoding task (i.e., the transparency judgments) was employed. This is because the meanings of whole words and constituents were being specifically accessed; thus, there was no contribution for conceptual fluency.

As mentioned, the retrieval orientation interpretation accounts for the presence and absence of the positivity associated with transparent correct rejections. Nevertheless, this interpretation is challenged by a finding in the present study. Assume that the positivity associated with the transparent words observed in Experiment 1 reflects a retrieval orientation effect. Would this suggest that the modulation of word transparency on the left parietal scalp region did not manifest the left parietal ERP old/new effect (an index of recollection-based recognition)? The results of Experiment 1 represent a combination of differences in the parietal effect and the retrieval orientation effect. As postulated by Rugg and Wilding (2000), the retrieval orientation effect should not be modulated by whether the test item is an old item or a new item. In other words, the retrieval orientation process revealed by the comparison of correct rejections for the opaque and transparent words should also have been engaged in the hits to the two types of words. In that case, the transparent hits should have been more positive than the opaque hits, demonstrating the same retrieval orientation effect on the correct rejections. In other words, negativity associated with the opaque hits as opposed to the transparent hits, reflecting the retrieval orientation effect, should have been detected. However, in Experiment 1, no differences were found between the two types of hits. A plausible interpretation is that the negative-going effect of retrieval orientation processing on the opaque hits was offset by the positive-going effect of recollection on the hits to the opaque words. In other words, the opaque words elicited a greater left parietal old/new effect than did the transparent words. However, this effect was coupled with the retrieval orientation effect.

11.6 Conclusion

The current study investigated the mechanisms underlying the recognition memory advantage of opaque words over transparent words by using ERP old/new effects. Similar to the findings of Han et al. (2014), the superior memory performance corresponding to opaque words induced a stronger left parietal effect, but not a stronger mid-frontal effect. These effects represent the contributions of recollection- and familiarity-based recognition, respectively. Notably, the advantage of the opaque words was reduced when the encoding task involved scrutinizing the meanings of the whole word and its constituents. Furthermore, the internal semantic structure of the transparent and opaque words likely elicited different retrieval orientations when the words were encoded with a superficial task (e.g., lexical decisions), but not when the words were encoded with a more semantically specific task (e.g., transparency judgments).

Acknowledgements This research was supported by grants from the Ministry of Science and Technology, Taiwan to Shih-kuen Cheng (MOST 111-2410-H-008-039, 109-2410-H-008-023).

References

Azimian-Faridani, N., & Wilding, E. L. (2006). The influence of criterion shifts on electrophysiological correlates of recognition memory. *Journal of Cognitive Neuroscience, 18*(7), 1075–1086. https://doi.org/10/bjnsh8

Chen, K.-J., Huang, C.-R., Chang, L.-P., & Hsu, H.-L. (1996). SINICA CORPUS : Design methodology for balanced corpora. *Proceedings of the 11th Pacific Asia Conference on Language, Information and Computation*, 167–176. http://hdl.handle.net/2065/12025

Cortese, M. J., Watson, J. M., Wang, J., & Fugett, A. (2004). Relating distinctive orthographic and phonological processes to episodic memory performance. *Memory and Cognition, 32*(4), 632–639. https://doi.org/10/b67btr

Curran, T. (2000). Brain potentials of recollection and familiarity. *Memory & Cognition, 28*(6), 923–938. https://doi.org/10.3758/Bf03209340

Delorme, A., & Makeig, S. (2004). EEGLAB: An open source toolbox for analysis of single-trial EEG dynamics including independent component analysis. *Journal of Neuroscience Methods, 134*(1), 9–21. https://doi.org/10/bqr2f2

Dobbins, I. G., Kroll, N. E. A., Yonelinas, A. P., & Liu, Q. (1998). Distinctiveness in recognition and free recall: The role of recollection in the rejection of the familiar. *Journal of Memory and Language, 38*(4), 381–400. https://doi.org/10/cbh9zp

Düzel, E., Yonelinas, A. P., Mangun, G. R., Heinze, H. J., & Tulving, E. (1997). Event-related brain potential correlates of two states of conscious awareness in memory. *Proceedings of the National Academy of Sciences of the United States of America, 94*(11), 5973–5978. https://doi.org/10/fn64v6

Fliessbach, K., Weis, S., Klaver, P., Elger, C. E., & Weber, B. (2006). The effect of word concreteness on recognition memory. *NeuroImage, 32*(3), 1413–1421. https://doi.org/10/cvtwcq

Gardiner, J. M., & Java, R. I. (1991). Forgetting in recognition memory with and without recollective experience. *Memory & Cognition, 19*(6), 617–623. https://doi.org/10.3758/BF03197157

Gardiner, J. M., & Parkin, A. J. (1990). Attention and recollective experience in recognition memory. *Memory & Cognition, 18*(6), 579–583. https://doi.org/10/cpk8qn

Gentner, D. (1981). Verb semantic structures in memory for sentences: Evidence for componential representation. *Cognitive Psychology, 13*(1), 56–83. https://doi.org/10/crpsq9

Glanc, G. A., & Greene, R. L. (2007). Orthographic neighborhood size effects in recognition memory. *Memory and Cognition, 35*(2), 365–371. https://doi.org/10/fdc9q7

Glanzer, M., & Adams, J. K. (1985). The mirror effect in recognition memory. *Memory & Cognition, 13*(1), 8–20. https://doi.org/10/fghx8q

Glanzer, M., & Adams, J. K. (1990). The mirror effect in recognition memory: Data and Theory. *Journal of Experimental Psychology: Learning, Memory, and Cognition, 16*(1), 5–16. https://doi.org/10/d3pxgs

Hamilton, M., & Rajaram, S. (2001). The concreteness effect in implicit and explicit memory tests. *Journal of Memory and Language, 44*(1), 96–117. https://doi.org/10/cjw4pj

Han, Y. J., Huang, S. C., Lee, C. Y., Kuo, W. J., & Cheng, S. K. (2014). The modulation of semantic transparency on the recognition memory for two-character Chinese words. *Memory and Cognition, 42*(8), 1315–1324. https://doi.org/10/gm9mqh

Heathcote, A., Ditton, E., & Mitchell, K. (2006). Word frequency and word likeness mirror effects in episodic recognition memory. *Memory and Cognition, 34*(4), 826–838. https://doi.org/10/bwxc8x

Herron, J. E., & Rugg, M. D. (2003). Strategic influences on recollection in the exclusion task: Electrophysiological evidence. *Psychonomic Bulletin & Review, 10*(3), 703–710. https://doi.org/10.3758/BF03196535

Hirshman, E., & Jackson, E. (1997). Distinctive perceptual processing and memory. *Journal of Memory and Language, 36*(1), 2–12. https://doi.org/10/bwdntb

Hornberger, M., Morcom, A. M., & Rugg, M. D. (2004). Neural correlates of retrieval orientation: Effects of study-test similarity. *Journal of Cognitive Neuroscience, 16*(7), 1196–1210. https://doi.org/10.1162/0898929041920450

Huang, C.-R., Ahrens, K., & Chen, K.-J. (1998). A data-driven approach to the mental Lexicon: Two studies on Chinese Corpus linguistics. *Bulletin of the Institute of History and Philology Academia Sinica, 69*(1), 151–180. https://doi.org/10/gm5c6v

Hunt, R. R., & Elliot, J. M. (1980). The role of nonsemantic information in memory: Orthographic distinctiveness effects on retention. *Journal of Experimental Psychology: General, 109*(1), 49–74. https://doi.org/10/dd4brv

Joordens, S., & Hockley, W. E. (2000). Recollection and familiarity through the looking glass: When old does not mirror new. *Journal of Experimental Psychology: Learning, Memory, and Cognition, 26*(6), 1534–1555. https://doi.org/10/cf3p3d

Kersten, A. W., & Earles, J. L. (2004). Semantic context influences memory for verbs more than memory for nouns. *Memory and Cognition, 32*(2), 198–211. https://doi.org/10/b7b7k5

Lee, H.-J., Cheng, S., Lee, C.-Y., & Kuo, W.-J. (2021). The neural basis of compound word processing revealed by varying semantic transparency and morphemic neighborhood size. *Brain and Language, 221*, Article 104985. https://doi.org/10.1016/j.bandl.2021.104985

Libben, G. (1998). Semantic transparency in the processing of compounds: Consequences for representation, processing, and impairment. *Brain and Language, 61*(1), 30–44. https://doi.org/10/dxg48h

Libben, G., Gibson, M., Yoon, Y. B., & Sandra, D. (2003). Compound fracture: The role of semantic transparency and morphological headedness. *Brain and Language, 84*(1), 50–64. https://doi.org/10/bff8tv

Lopez-Calderon, J., & Luck, S. J. (2014). ERPLAB: An open-source toolbox for the analysis of event related potentials. *Frontiers in Human Neuroscience, 8*. https://doi.org/10/gdz7rc

MacGregor, L. J., & Shtyrov, Y. (2013). Multiple routes for compound word processing in the brain: Evidence from EEG. *Brain and Language, 126*(2), 217–229. https://doi.org/10.1016/j.bandl.2013.04.002

McCarthy, G., & Wood, C. C. (1985). Scalp distributions of event-related potentials—An ambiguity associated with analysis of variance models. *Electroencephalography and Clinical Neurophysiology, 61*(3), S226–S227. https://doi.org/10/csf576

Migo, E. M., Mayes, A. R., & Montaldi, D. (2012). Measuring recollection and familiarity: Improving the remember/know procedure. *Consciousness and Cognition, 21*(3), 1435–1455. https://doi.org/10/f357jj

Morcom, A. M. (2015). Resisting false recognition: An ERP study of lure discrimination. *Brain Research, 1624*, 336–348. https://doi.org/10.1016/j.brainres.2015.07.049

Morris, C. D., Bransford, J. D., & Franks, J. J. (1977). Levels of processing versus transfer appropriate processing. *Journal of Verbal Learning and Verbal Behavior, 16*(5), 519–533. https://doi.org/10/cjf38v

Paller, K. A., Voss, J. L., & Boehm, S. G. (2007). Validating neural correlates of familiarity. *Trends in Cognitive Sciences, 11*(6), 243–250. https://doi.org/10/brpqv8

Rajaram, S. (1996). Perceptual effects on remembering: Recollective processes in picture recognition memory. *Journal of Experimental Psychology: Learning Memory and Cognition, 22*(2), 365–377. https://doi.org/10/crp35p

Rotello, C. M., & Heit, E. (1999). Two-Process models of recognition memory: Evidence for recall-to-reject? *Journal of Memory and Language, 40*(3), 432–453. https://doi.org/10.1006/jmla.1998.2623

Rugg, M. D., Allan, K., & Birch, C. S. (2000). Electrophysiological evidence for the modulation of retrieval orientation by depth of study processing. *Journal of Cognitive Neuroscience, 12*(4), 664–678. https://doi.org/10.1162/089892900562291

Rugg, M. D., & Curran, T. (2007). Event-related potentials and recognition memory. *Trends in Cognitive Sciences, 11*(6), 251–257. https://doi.org/10/bgfrb5

Rugg, M. D., Fletcher, P. C., Allan, K., Frith, C. D., Frackowiak, R. S. J., & Dolan, R. J. (1998a). Neural correlates of memory retrieval during recognition memory and cued recall. *NeuroImage, 8*(3), 262–273. https://doi.org/10/c78jbm

Rugg, M. D., Schloerscheidt, A. M., & Mark, R. E. (1998b). An electrophysiological comparison of two indices of recollection. *Journal of Memory and Language, 39*(1), 47–69. https://doi.org/10.1006/jmla.1997.2555

Rugg, M. D., & Wilding, E. L. (2000). Retrieval processing and episodic memory. *Trends in Cognitive Sciences, 4*(3), 108–115. https://doi.org/10.1016/S1364-6613(00)01445-5

Stretch, V., & Wixted, J. T. (1998). On the difference between strength-based and frequency-based mirror effects in recognition memory. *Journal of Experimental Psychology: Learning, Memory, and Cognition, 24*(6), 1379–1396. https://doi.org/10.1037/0278-7393.24.6.1379

Tsai, J.-L., Lee, C.-Y., Lin, Y.-C., Tzeng, O. J. L., & Hung, D. L. (2006). Neighborhood size effects of Chinese words in Lexical decision and reading. *Language and Linguistics, 7*(3), 659–675.

Tulving, E. (1984). Précis of Elements of episodic memory. *Behavioral and Brain Sciences, 7*(2), 223–238. https://doi.org/10/cfj88t

Tulving, E., & Thomson, D. M. (1973). Encoding specificity and retrieval processes in episodic memory. *Psychological Review, 80*(5), 352–373. https://doi.org/10/cgj2rr

Wilding, E. L., & Rugg, M. D. (1996). An event-related potential study of recognition memory with and without retrieval of source. *Brain, 119*(3), 889–905. https://doi.org/10/bkf8mt

Wixted, J. T., & Mickes, L. (2010). A continuous dual-process model of remember/know judgments. *Psychological Review, 117*(4), 1025–1054. https://doi.org/10/cnp78j

Woodruff, C. C., Hayama, H. R., & Rugg, M. D. (2006). Electrophysiological dissociation of the neural correlates of recollection and familiarity. *Brain Research, 1100*, 125–135. https://doi.org/10/fbv63v

Yonelinas, A. P. (1994). Receiver-operating characteristics in recognition memory: Evidence for a dual-process model. *Journal of Experimental Psychology. Learning, Memory, and Cognition, 20*(6), 1341–1354. https://doi.org/10.1037//0278-7393.20.6.1341

Yonelinas, A. P., & Jacoby, L. L. (1995). The relation between remembering and knowing as bases for recognition: Effects of Size Congruency. *Journal of Memory and Language, 34*(5), 622–643. https://doi.org/10/b3gq9t

Chapter 12
The Application of Information Theory in Studying Motor Control

Hsin-Ping Tien and Erik Chihhung Chang

Abstract Being one of the pillars of the cognitive revolution in psychology, information theory provides valuable insights into how the brain functions as an information system in various cognitive functions. In motor control, the body serves as a conduit for information transmission between the mind and the physical world. With this approach, researchers have demonstrated how optimization of information processing enhances movement efficiency and accuracy, with far-reaching impacts on robotics and AI in terms of replicating human motor learning and control. Overall, the information-theoretic perspective provides a comprehensive evaluation framework that allows disparate cognitive functions to be studied under a unified, information-centric paradigm.

12.1 Information Theoretic Framework: An Integrative Perspective on Cognition

The information-theoretic framework has been embraced across various cognitive domains, such as language and motor functions, with research efforts striving to integrate these complex systems within a unified theoretical perspective. This approach, predicated on Shannon's (1948) foundational work in information theory, interprets the brain as a sophisticated information processor, deciphering cognitive processes through the lens of entropy, uncertainty, and predictive computations (Friston, 2010). Such a framework proposes a cohesive mathematical language that links multiple levels of cognitive analysis, from the intricacies of neural circuitry to the overarching algorithms, culminating in an integrated understanding of behavior (Piccinini & Scarantino, 2011). This chapter starts by introducing fundamental concepts of the

H.-P. Tien · E. C. Chang (✉)
Institute of Cognitive Neuroscience, National Central University, Taoyuan, Taiwan
e-mail: audachang@gmail.com

H.-P. Tien
Taiwan International Graduate Program in Interdisciplinary Neuroscience, National Central University and Academia Sinica, Taipei, Taiwan

D. H. Wu (ed.), *Cognitive and Neural Foundations of Chinese Reading*, Chinese Language Learning Sciences, https://doi.org/10.1007/978-981-96-6669-0_12

information-theoretic approach, followed by its application in studies of motor control, and finally, an example demonstrating how it unifies the assessment of motor control and higher order cognitive functions. Through this brief introduction, we hope to attract more attention to the application of the information-theoretic approach in profiling individuals' cognitive functions meaningfully and quantitatively from an integrative perspective.

The information-theoretic approach refers to analyzing and modeling phenomena using concepts and tools from information theory. The key aspects of the information theoretic approach include:

- Viewing systems and processes in terms of information sources, channels, signals, noise, coding, and communication. This involves representing the system elements mathematically in terms of random variables, probability distributions, entropy, etc.
- Quantifying the amount of information in terms of entropy, mutual information, relative entropy, channel capacity, etc. This allows measuring and comparing the information content of different systems.
- Modeling statistical dependencies and structures in the system using joint and conditional entropies. This captures how the underlying relationships affect the flow of information.
- Looking at optimization of information transmission, such as data compression. Information theory provides fundamental limits and optimal coding schemes.
- Evaluating information loss or distortion due to noise. Measuring the robustness and reliability of communication through the channel.
- Leveraging the mathematical framework of information theory to quantify and analyze the informational aspects of the system or process. This complements physical or mechanistic models.

In its essence, the information-theoretic approach provides an abstract, probabilistic framework to study fundamental questions about information and communication properties of systems. It shifts traditional perspectives of various research fields to focus on signals, uncertainties, and statistical relationships transcending the physical details. This approach has been highly influential in fields like telecommunications, data compression, linguistics, and neuroscience. Here, we will use motor control as an example of applying the information-theoretic framework in experimental cognitive psychology.

12.2 Fitts' Law: A New Look at Motor Control Through Information Theory

In a groundbreaking paper, Fitts (1954) proposed a computational model based on observations of human performance in a manual aiming task (Fig. 1A), positing that the movement time (MT) remains consistent when the ratio between movement

amplitude (A) and target width (W) is constant. This implies that longer movements aimed at broader targets take approximately the same amount of time as shorter movements aimed at narrower targets. Furthermore, Fitts observed that the MT escalates when the A to W ratio increases, which can occur through enlarging A, reducing W, or a combination of both. These observations were integrated into a unified equation by Fitts in the following form of the Index of Difficulty
(ID):

$$ID = \log_2 \frac{2A}{W} \text{ bits}$$

Here, A stands for the amplitude of the movement, and W represents the *precision tolerance*, like the width of the target. This model links the execution of movements to the process of transmitting information. In this framework, movements are quantified by ID. The human motor system, in performing a movement task, is conceptualized as transmitting a certain number of "bits of information." By dividing the total bits by the time taken for the movement, one can determine a transmission rate in "bits per second."

Following the assessment of information associated with a particular movement, Fitts introduced the Index of Performance (I_p) to measure information capacity by dividing the ID by the movement time (MT), which is equivalent to the slope of the linear function between MT and ID (Fig. 1B). Fitts' model is conceptually linked to Hick's law in speed-reaction time experiments (Hick, 1952a), which suggests that an increase in the number of alternatives leads to higher uncertainty and entails more information. Due to limited information capacity, processing responses with higher loading of information contents will take more time.

Fitts' original information calculation can result in zero or even negative information values. Specifically, zero information is elicited when the amplitude of movement (A) is equal to half the precision tolerance *(W/2)*, while negative information

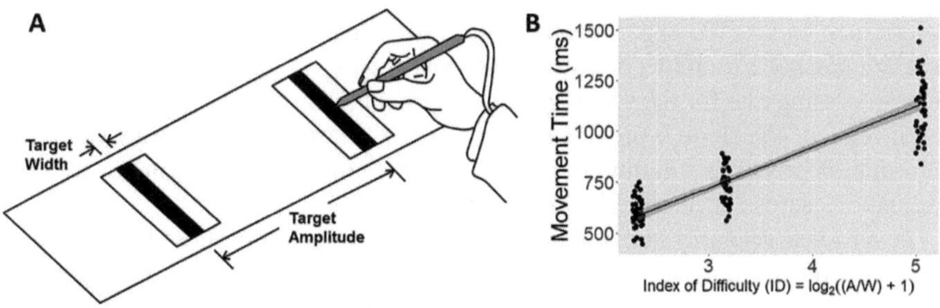

Fig. 12.1 Fitts' paradigm and typical outcomes. A. Fitts' original experimental settings of the manual tapping task (Fitts, 1954); B. Typical outcomes of movement time as a function of the index of difficulty. Each dot represents the average movement time of a participant under the ID designated on the x-axis

is generated when W exceeds $A/2$. In response to these nuances, several modifications of Fitts' law have been proposed to enhance the information provided by aiming movements. For instance, Welford et al. (1969) introduced the concept of the "landing procedure" for movements approaching the edge of the target position. They suggested that movement terminations should be concentrated in the center of the target, leading to a modified *ID* formula:

$$\log_2 \frac{A + \frac{1}{2}W}{W}$$

Additionally, an analogy with Shannon's Theorem 17 (Shannon, 1948) was incorporated to quantify the *ID* in a different way, as proposed by Mackenzie (1992):

$$\log_2 \frac{A + W}{W}$$

Mackenzie's formulation of *ID, derived from Shannon's Theorem 17,* has been noted to provide better performance when the signal-to-noise ratio is low. In contrast, Fitts' *original ID* formulation is more suitable when the signal amplitude significantly exceeds the noise level (MacKenzie, 1989). Importantly, Mackenzie's *ID* avoids the generation of negative information when describing relatively high precision tolerance movements. These refinements in ID calculations allow for a more accurate assessment of information capacity in different motor control contexts.

I_p, the *ID* to MT ratio, is a critical metric used to gauge the information processing rate during aiming movements, measured in bits per second. This measure is based on the central assumption that information capacity remains relatively constant across various movement parameters.

However, empirical observations have shown that I_p is not consistent across different *ID* values within the same type of motor task (Zhai, 2004). To further explore the relationship between MT and *ID*, researchers have employed a linear regression model, expressed as $MT = a + bID$. The parameters a and b can be determined by specific experimental motor task. What is particularly noteworthy is that $1/b$ has been found to be more consistent than I_p in quantifying the information processing rate for the same type of movement (Zhai, 2004). This suggests that the inverse of the slope b provides a more stable measure for assessing the rate of information processing in motor tasks, allowing for more accurate comparisons and evaluations across different conditions and contexts.

In an ideal scenario, the y-intercept of Fitts' law should be zero, implying a predicted MT of 0 ms for a task with an ID of 0. This intuitive expectation arises from the notion that a movement with ID = 0 should theoretically require no information to process. However, the persistence of a non-zero y-intercept in Fitts' law posed a challenge from its inception (Fitts & Peterson, 1964; Soukoreff & MacKenzie, 2004). In certain studies, negative y-intercept values have even been observed (Fitts & Peterson, 1964; Kantowitz & Elvers, 1988). Fitts acknowledged the physical impossibility of a zero or negative MT and asserted that the y-intercept should theoretically always be positive (Fitts & Radford, 1966). Various interpretations of non-zero y-intercepts include inherent delays in the psychomotor system (Fitts & Radford, 1966) and uncontrollable muscle activity at the initiation or conclusion of the movement task (Mackenzie, 1992).

Over the years, since its initial introduction by Fitts, this law has become one of the most enduring and frequently cited models in experimental psychology. Its applicability and validity have been consistently confirmed through psychomotor studies in various environments, ranging from microscopic to aquatic settings. These studies have repeatedly found strong correlations between the index of difficulty, as per Fitts' model, and the time required to complete a movement task. The fields of kinematics and human factors have been particularly enriched by research utilizing Fitts' analogy to explore aspects of human performance.

12.3 Factors Modulating Parameters of Fitts' Law

Based on the premise that information capacity varies with specific motor control processes, different motor-related factors are likely to modulate this capacity in the context of Fitts' law. Fitts noted that the characteristics of the movement and tool properties have a minimal impact on information processing (Fitts, 1954). However, Radwin et al. (1990) found that both the direction of movement and the executing effector can influence the Index of Difficulty (ID).

Furthermore, research indicates that movements involving the fingers and wrist over short distances have higher information processing rates (measured in bits/sec) compared to longer arm movements. Langolf et al. (1976) attributed this difference to the engagement of different limb segments. This variation is also explainable through the concept of degrees of freedom in movement, which can increase the steepness of Fitts' law's slope for more distant movements. The complexity of movements, related to degrees of freedom, resonates with Hick's law. Yet, Bertucco et al., (2013) observed no changes in the slope of Fitts' law when manipulating trunk rotation angles. Instead, they noted alterations in the intercept, which might be attributed to Coriolis forces.

Kinetic factors have also been examined in the context of Fitts' law. Loaded movements have been found not modulating the slope of Fitts' law but affecting the intercept (Bonnetblanc, 2008). These kinetic factors can involve various aspects of movement control, such as elbow angle, acceleration/deceleration duration, peak velocity, and EMG responses. However, these kinetic constraints unrelated to the complexity of movement may not affect the information capacity. Also, sensorimotor experience of tolerance levels has been shown to modulate the intercept but not the slope of Fitts' law (Tang et al., 2018). Interestingly, changes in the amplitude do not appear to affect the intercept and slope of the Fitts' law equation. The intercept modulation of kinetic constraints and sensorimotor experience suggests that the "internal" noise level adopted by the subject can be modulated by experience and may further alter the *ID*. On the other hand, the degree of freedom in movement could modulate the information capacity of movement.

12.4 Dynamic Feedback and Feedforward Information in Online Movement Control

In visuomotor tracking tasks involving visual input (X) and motor output (Y), the use of mutual information to determine the information processing rate is key. This approach includes defining delays and historical depth. Transfer entropy, a critical tool in this context, models the information processed through conditional mutual information, balancing recent output with previous inputs, given the output's history. Lam and Zénon (2021) highlighted the pivotal role of predictability in influencing the information processing rates of feedback and feedforward systems, noting that these effects are inversely related. This insight underscores the necessity for future research to not only quantify motor control information capacity but also dissect the distinct contributions of feedback and feedforward information over time. Such studies should take into account the cumulative impact of visuomotor delay, moving beyond a mere aggregation of motor planning information.

12.5 Sensory Information Before and During the Movement

The loci of Fitts' law have intrigued researchers: Does its focus lie in initial information acquisition and transmission or in online control aspects? Wu et al. (2010) separated motor control into stages like cognition, planning, feedback, and control. They showed that visual feedback impacts the slope relating MT and ID. Limiting feedback increased MT variations and reduced accuracy, highlighting the visual feedback's role. While valuable, Fitts' law may not fully capture online control complexity. In visuomotor tracking tasks, for example, mutual information is key, encompassing delays and history. The finite capacity of human control, bounded by noise, implies an upper limit on processing rate. This relates to the speed-accuracy trade-off, where faster movements enable lower precision.

While Fitts' Law provides a valuable perspective on the initial stages of information processing, it may not fully capture the complex mechanisms of online control. For example, in visuomotor tracking tasks, where visual input (X) aids in motor output (Y), mutual information is crucial for quantifying the information processing rate, including defined delays and history depth. The human visuomotor control system's finite capacity, bounded by noise, mirrors an upper limit in information processing rate. This finite capacity relates to the observed speed-accuracy trade-off in motor tasks, where increased speed (with higher noise tolerance) results in less precision, and vice versa.

Gori et al. (2018) advocated for clearer definitions of capacity and throughput in Fitts' law contexts, enhancing previous interpretations and incorporating concepts like misses at rate ε. This model focuses on minimizing movement time rather than average performance (Gori et al., 2017). Capacity is defined as the maximum mutual information *I(X;Y)* transmitted through the channel, ensuring reliable data transmission. *Throughput*, the effective data transmission speed, is defined as MT_i / ID_i in motor tasks, representing effective information transmission speed. Soukoreff and MacKenzie (2009) introduced the concept of *equivocation*, namely the difference between throughput and source entropy, to model the attainable region on source entropy to equivocation plots. The source entropy processing range, limited by channel capacity, expands with increased precision tolerance, aligning with the speed-accuracy trade-off.

Previous studies focused on the maximum capacity constrained to the Eh line in source entropy to *equivocation plots*. Soukoreff and MacKenzie (2009) noted human motor performance often deviates from ideal information transmission lines, introducing a constant (α) to modulate the human-achievable equivocation: $\mathrm{Eh} = \alpha(\mathrm{I} - \mathrm{C})$. This approach, using the best MT to infer Fitts' law, takes into account field study variations in MT. Gori et al. (2017) suggested distinguishing between *deviation of movement termination* (spatial error) and *misses* (*binary hit-or-miss indicator*, i.e., *erasure*), and proposed a new effective index ID(ε): $(1 - \varepsilon)\log_2 \frac{A+W}{W}$, where ε indicates the target miss rate. The intricate measurements and integration of *source entropy*, *information capacity*, *throughput*, and *equivocation* enriches the understanding of the nature of human motor performance.

12.6 Bridging Motor Skills and Cognition: An Information-Theoretic Nexus

While information theory is foundational to the "cognitive revolution" as highlighted by Neisser in 1976, its application in cognitive psychology has often been more metaphorical than quantitative. Typically, cognitive psychology experiments do not directly quantify the amount of information in stimuli in relation to cognitive processing. However, an information theory-based approach can provide a straightforward and precise means of measuring cognitive functions. Take, for instance, the assessment of executive function using tasks that involve processing uncertainty, such the Majority Function Task (MFT; Fan, 2014; Fan et al., 2008, 2014; Wang et al., 2011). In MFT, participants are presented with arrays of arrows pointing left or right and must identify the predominant direction. The task's complexity is varied by adjusting the number of arrows and their arrangement (congruency). The researchers hypothesize different search strategies, like exhaustive, self-terminating, or grouping searches. This task's computational demand for each trial is quantified in bits of entropy. Fan et al. (2008) found that the grouping search strategy most effectively represented the linear relationship between reaction time (RT) and computational load in MFT, where RT is expressed as a function of the average number of arrows scanned per trial (s): $RT = a + b\log 2(s)$. Analogous to the Fitts' law, here the slope (**b**) represents the additional processing time required per bit of increased load, indicating processing efficiency. On the other hand, the intercept (**a**) corresponds to the processing time for the simplest task possible (0 bits), akin to the reaction time for responding to a single arrow. Thus, information theory provides a precise means to measure cognition in uncertain tasks like MFT.

The parallels between the MFT and Fitts' Law are evident, as both exhibit a linear relationship between a measure of speed and the quantity of information processed. This analogy extends to a comparison made by Sleimen-Malkoun et al. (2013) between Fitts' law and Hick-Hyman's law in young and elderly groups. Hick-Hyman's Law, in contrast to Fitts' Law, characterizes the choice reaction time (CRT) as a linear function of the information entropy in response selection, essentially the binary logarithm of potential stimulus–response associations, while maintaining the motor response complexity at its minimum (Hawkins et al., 2012; Hick, 1952b; Hülür et al., 2015; Hyman, 1953). The study revealed that, although the slopes of both laws were indistinguishable in the elderly, younger participants exhibited steeper slopes in Fitts' Law compared to Hick-Hyman's Law.

Following the same vein, Chang (2021) compared aging across various cognitive domains using experimental designs that manipulate computational load within a consistent informational framework. To quantitatively assess these designs, linear functions for the MFT and Fitts' Law were evaluated using the Generalized Linear Mixed Model (GLMM), which examined the interplay of age, task, and computational load both at a collective level (as a fixed effect) and on an individual basis (as a random effect). The outcomes indicated that among the elderly, increasing computational load led to similar rates of increment in movement time (MT) and reaction time (RT). In contrast, younger participants exhibited a steeper RT slope in the MFT compared to the MT slope in Fitts' Law (Fig. 12.2). Notably, stronger correlations were observed in the elderly group between individual slope and intercept estimates across and within tasks, compared to the younger group. In addition, the observed stronger cross-domain correlations further corroborate the dedifferentiation hypothesis. This research approach facilitates more direct cross-domain comparisons than would be possible with different tasks that do not align in computational load. Without this alignment, comparisons would rely on summarized and transformed data, potentially reducing statistical power and limiting inferences about lifespan trajectories across cognitive domains.

Synthesizing the findings from studies that compare motor and cognitive functions within the information-theoretic framework, there is support for the cognitive aging dedifferentiation hypothesis. This hypothesis posits that with age-related reductions in cognitive resources, distinct cognitive processes that are separated in youth tend to converge and use shared resources in older age, leading to increased similarity (Baltes & Lindenberger, 1997; Lindenberger & Baltes, 1994). However, it's important to note that the evidence supporting the dedifferentiation hypothesis has been mixed

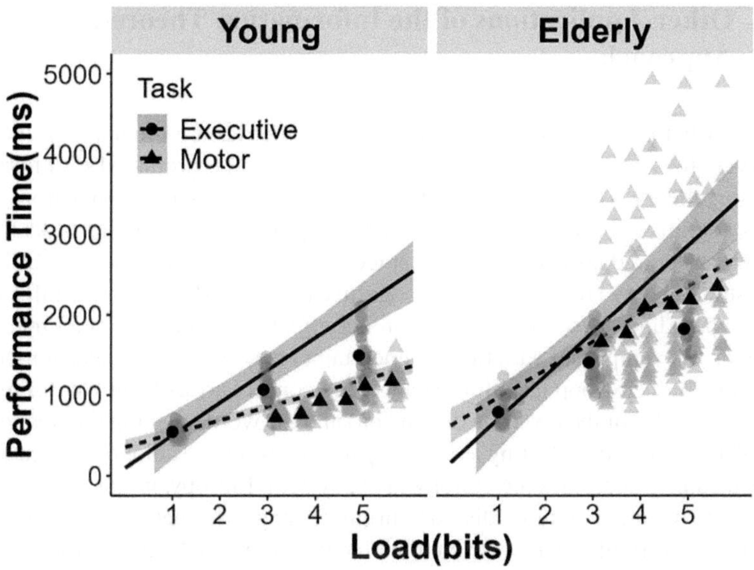

Fig. 12.2 Predicted values of performance time as a function of Age Group and Task. The solid line (pinkish shade) and dashed line (cyan shade) indicate the linear function (95% CI) between performance time and Load in the MFT (executive) and Fitts (motor) task, respectively. The dot and triangle symbols represent the group mean of the analysed data at each combination between Task and Load in the experiment, with error bars indicating the 95% CI normalized for within-subject error (adapted from Chang [2021])

(Hülür et al., 2015), potentially influenced by research design (longitudinal vs. cross-sectional) and the nature of the sample (including life-span cohorts with young and middle-aged participants or focusing solely on the elderly).

Besides these factors, a third potential influence arises from the application of distinct methods used to measure and analyze different cognitive functions, which might not be comparable and can significantly affect the results and interpretations of aging trends. La Fleur et al. (2018) illustrated this point by showing that including a speed of processing test in correlation analyses with other abilities influenced the tendency for linear dedifferentiation among cognitive domains. Since studies on dedifferentiation typically rely on distinct methods to quantify various cognitive domains, the resultant analyses and interpretations are often based on transformed scores. Thus, considering the current study's findings, along with those from previous reports, it appears that some degree of variation in the inconsistencies across studies might be attributable to distortions in measurement resulting from the transformation and averaging of performance indices. To this end, designing cognitive tasks and measuring performance under the information-theoretic framework may provide a fair perspective for cross-domain comparison.

12.7 Other Applications of the Information Theoretic Approach

It is somewhat paradoxical that core information metrics are often absent in cognitive psychology experiments, a field that professes to be grounded in information processing theory. The adoption of information-entropy-based manipulations across various cognitive abilities might necessitate a significant redesign of tasks that are currently structured around widely accepted operational definitions. Fan (2014) has demonstrated how to reframe existing cognitive control tasks within the information theoretic paradigm. We can interpret conflict effects stemming from an interfering dimension as examples of variations in uncertainty between conflict and non-conflict conditions. In the Stroop task, for instance, the conflict arises from the discrepancy between the color of the text (one dimension) and the word meaning (another dimension). This conflict, manifesting as a discrepancy between the color of the text and the word's meaning, increases the entropy in the system. Entropy, in this context, refers to the level of uncertainty or disorder in processing information. When the brain encounters a word like "RED" printed in blue ink, it faces a higher entropy situation due to the incongruence between the color and the word's meaning. Resolving this incongruence requires the brain to suppress the automatic reading response (a lower entropy task) in favor of the more complex task of naming the ink color (a higher entropy task). This increased cognitive effort due to higher entropy is reflected in slower response times and a greater likelihood of errors, illustrating the impact of conflicting information dimensions in cognitive processing.

Another good example is the flanker task. Through an information-theoretic lens, the conflict in the flanker task arises from the presence of distracting flankers around a central target introduces increased entropy, signifying higher uncertainty in the cognitive process. When participants encounter incongruent flankers, the brain has to manage conflicting streams of information—the direction of the central arrow (target) and the surrounding arrows (flankers). This situation heightens the cognitive load due to the need to suppress responses to the misleading flanker information. Consequently, the increased entropy manifests as slower reaction times and a higher error rate in incongruent conditions, compared to congruent ones where flanker and target information align, resulting in lower entropy and easier cognitive processing.

Moreover, one could quantify uncertainty using surprise and entropy calculations in paradigms that investigate phenomena like the oddball effect, Go/No-Go tasks, or task switching, which all involve contrasting responses to events of varying probabilities. For example, in the task-switching paradigm, the cognitive challenge arises from the need to switch between different tasks, each with its own set of rules and responses. This switch introduces high entropy—an increase in uncertainty and cognitive load—as the brain must inhibit the rules and responses of the previous task while activating those of the new task. This increased entropy is evidenced by the typically observed switch cost, manifested as slower response times and higher error rates when switching tasks compared to repeating the same task. This reflects the brain's effort to reconfigure and stabilize a new cognitive state, aligning with principles of information theory.

The information-theoretic paradigm also shows promise for elucidating mechanisms of language processing. For instance, experiments could manipulate linguistic uncertainty by varying lexical frequency and syntactical complexity in sentences and passages (Hale, 2001). Increased surprisal and entropy from rare words and convoluted syntax require heightened cognitive effort, measurable via slower reading times, more errors in comprehension questions, and modulated neurological signals (Levy, 2008). Moreover, analyzing linguistic inputs and outputs in terms of transmitted bits of information per unit time provides a quantitative basis to compare the efficiency of language processing across tasks and populations (Jaeger, 2010). An information-centric framework centered on quantifying transmission, storage, and loss of linguistic information at multiple levels ranging from phonemes to semantics could yield fresh insights into the nature of the language system (Linzen & Jaeger, 2015). Going further, some researchers envision the development of a "calculus of thought" founded on information theoretic formulations, providing an integrated perspective on cognition encompassing perception, executive functions, memory, language, and more under a unified set of information processing principles (Piccinini & Scarantino, 2011). Applying this approach to complex domains like long-term memory and language is challenging initially. However, as the information framework expands to encompass basic cognitive processes, it may evolve into a comprehensive system that explicates cognitive functions regarding information exchange and storage (Wolff, 1982). This unified analysis under common information metrics could transform understanding of cognition across domains (Dell, 1986). Indeed, the information paradigm heralds new horizons for cognitive science, propelling more precise and integrated models of the human mind.

References

Baltes, P. B., & Lindenberger, U. (1997). Emergence of a powerful connection between sensory and cognitive functions across the adult life span: A new window to the study of cognitive aging? *Psychology and Aging, 12*(1), 12–21. https://doi.org/10.1037/0882-7974.12.1.12

Bertucco, M., Cesari, P., & Latash, M. L. (2013). Fitts' Law in early postural adjustments. *Neuroscience, 231*, 61–69. https://doi.org/10.1016/j.neuroscience.2012.11.043

Bonnetblanc, F. (2008). Pointing beyond reach: The slope of Fitts's law increases with the introduction of new effectors independently of kinetic constraints. *Motor Control, 12*(1), 38–54. https://doi.org/10.1123/mcj.12.1.38

Chang, E. C. (2021). Information-theoretic quantification of dedifferentiation in the aging of motor and executive functions. In *Frontiers in Aging Neuroscience* (Vol. 13): Frontiers Media SA.

Dell, G. S. (1986). A spreading-activation theory of retrieval in sentence production. *Psychological Review, 93*(3), 283–321. https://doi.org/10.1037/0033-295x.93.3.283

Fan, J. (2014). An information theory account of cognitive control. *Frontiers in Human Neuroscience, 8*, 1–16. https://doi.org/10.3389/fnhum.2014.00680

Fan, J., Guise, K. G., Liu, X., & Wang, H. (2008). Searching for the majority: Algorithms of voluntary control. *PLoS ONE, 3*, Article e3522. https://doi.org/10.1371/journal.pone.0003522

Fan, J., Van Dam, N. T., Gu, X., Liu, X., Wang, H., Tang, C. Y., & Hof, P. R. (2014). Quantitative characterization of functional anatomical contributions to cognitive control under uncertainty. *Journal of Cognitive Neuroscience, 26*, 1490–1506. https://doi.org/10.1162/jocn_a_00554

Fitts, P. M. (1954). The information capacity of the human motor system in controlling the amplitude of movement. *Journal of Experimental Psychology, 47*(6), 381–391. https://doi.org/10.1037/h0055392

Fitts, P. M., & Peterson, J. R. (1964). Information capacity of discrete motor responses. *Journal of Experimental Psychology, 67*(2), 103–112. https://doi.org/10.1037/h0045689

Fitts, P. M., & Radford, B. K. (1966). Information capacity of discrete motor responses under different cognitive sets. *Journal of Experimental Psychology, 71*(4), 475–482. https://doi.org/10.1037/h0022970

Friston, K. (2010). The free-energy principle: A unified brain theory? *Nature Reviews Neuroscience, 11*, 127–138.

Gori, J., Rioul, O., & Guiard, Y. (2018). Speed-accuracy tradeoff: A Formal Information-Theoretic Transmission Scheme (FITTS). *ACM Transactions on Computer-Human Interaction, 25*(5), Article 27. https://doi.org/10.1145/3231595

Gori, J., Rioul, O., Guiard, Y., & Beaudouin-Lafon, M. (2017). One Fitts' law, two metrics. In R. Bernhaupt, G. Dalvi, A. Joshi, D. K. Balkrishan, J. O'Neill, & M. Winckler (Eds.), *Human-Computer Interaction—INTERACT 2017*.

Hale, J. (2001). *A probabilistic earley parser as a psycholinguistic model*. Second meeting of the North American Chapter of the Association for Computational Linguistics on Language technologies 2001—NAACL '01, https://doi.org/10.3115/1073336.1073357

Hawkins, G. E., Brown, S. D., Steyvers, M., & Wagenmakers, E. J. (2012). An optimal adjustment procedure to minimize experiment time in decisions with multiple alternatives. *Psychonomic Bulletin & Review, 19*(2), 339–348. https://doi.org/10.3758/s13423-012-0216-z

Hick, W. E. (1952a). On the rate of gain of information. *The Quarterly Journal of Experimental Psychology, 4*, 11–26. https://doi.org/10.1080/17470215208416600

Hick, W. E. (1952b). On the rate of gain of information. *Quarterly Journal of Experimental Psychology, 4*, 11–26. https://doi.org/10.1080/17470215208416600

Hülür, G., Ram, N., Willis, S. L., Schaie, K. W., & Gerstorf, D. (2015). Cognitive dedifferentiation with increasing age and proximity of death: Within-person evidence from the Seattle Longitudinal Study. *Psychology and Aging, 30*, 311–323. https://doi.org/10.1037/a0039260

Hyman, R. (1953). Stimulus information as a determinant of reaction time. *Journal of Experimental Psychology, 45*(3), 188.

Jaeger, T. F. (2010). Redundancy and reduction: Speakers manage syntactic information density. *Cognitive Psychology, 61*(1), 23–62. https://doi.org/10.1016/j.cogpsych.2010.02.002

Kantowitz, B. H., & Elvers, G. C. (1988). Fitts' law with an isometric controller: Effects of order of control and control-display gain. *Journal of Motor Behavior, 20*(1), 53–66. https://doi.org/10.1080/00222895.1988.10735432

La Fleur, C. G., Meyer, M. J., & Dodson, C. (2018). Exploring dedifferentiation across the adult lifespan. *Psychology and Aging*. https://doi.org/10.1037/pag0000274

Lam, S. Y., & Zénon, A. (2021). Information rate in humans during visuomotor tracking. *Entropy (Basel), 23*(2). https://doi.org/10.3390/e23020228

Langolf, G. D., Chaffin, D. B., & Foulke, J. A. (1976). An investigation of fitts' law using a wide range of movement amplitudes. *Journal of Motor Behavior, 8*(2), 113–128. https://doi.org/10.1080/00222895.1976.10735061

Levy, R. (2008). Expectation-based syntactic comprehension. *Cognition, 106*(3), 1126–1177. https://doi.org/10.1016/j.cognition.2007.05.006

Lindenberger, U., & Baltes, P. B. (1994). Sensory functioning and intelligence in old-age—A strong connection. *Psychology and Aging, 9*(3), 339–355. https://doi.org/10.1037/0882-7974.9.3.339

Linzen, T., & Jaeger, T. F. (2015). Uncertainty and expectation in sentence processing: Evidence from subcategorization distributions. *Cognitive Science, 40*(6), 1382–1411. https://doi.org/10.1111/cogs.12274

MacKenzie, I. S. (1989). A note on the information-theoretic basis of Fitts' law. *Journal of Motor Behavior, 21*(3), 323–330. https://doi.org/10.1080/00222895.1989.10735486

Mackenzie, I. S. (1992). *Fitts' law as a performance model in human-computer interaction.* University of Toronto.

Piccinini, G., & Scarantino, A. (2011). Information processing, computation, and cognition. *Journal of Biological Physics, 37*(1), 1–38. https://doi.org/10.1007/s10867-010-9195-3

Radwin, R. G., Vanderheiden, G. C., & Lin, M.-L. (1990). A method for evaluating head-controlled computer input devices using Fitts' Law. *Human Factors, 32*(4), 423–438. https://doi.org/10.1177/001872089003200405

Shannon, C. E. (1948). A mathematical theory of communication. *Bell System Technical Journal, 27*(3), 379–423. https://doi.org/10.1002/j.1538-7305.1948.tb01338.x

Sleimen-Malkoun, R., Temprado, J. J., & Berton, E. (2013). Age-related dedifferentiation of cognitive and motor slowing: Insight from the comparison of Hick-Hyman and Fitts' laws. *Frontiers in Aging Neuroscience, 5*, 1–12. https://doi.org/10.3389/fnagi.2013.00062

Soukoreff, R. W., & MacKenzie, I. S. (2004). Towards a standard for pointing device evaluation, perspectives on 27 years of Fitts' law research in HCI. *International Journal of Human-Computer Studies, 61*(6), 751–789. https://doi.org/10.1016/j.ijhcs.2004.09.001

Soukoreff, R. W., & MacKenzie, I. S. (2009, October 11–14). *An informatic rationale for the speed-accuracy trade-off*. 2009 IEEE International Conference on Systems, Man and Cybernetics.

Tang, R., Shen, B., Sang, Z., Song, A., & Goodale, M. A. (2018). Fitts' Law is modulated by movement history. *Psychonomic Bulletin & Review, 25*(5), 1833–1839. https://doi.org/10.3758/s13423-017-1367-8

Wang, H., Liu, X., & Fan, J. (2011). Cognitive control in majority search: A computational modeling approach. *Frontiers in Human Neuroscience, 5*, 16. https://doi.org/10.3389/fnhum.2011.00016

Welford, A. T., Norris, A. H., & Shock, N. W. (1969). Speed and accuracy of movement and their changes with age. *Acta Psychologica, 30*, 3–15. https://doi.org/10.1016/0001-6918(69)90034-1

Wolff, J. G. (1982). Language acquisition, data compression and generalization. *Language & Communication, 2*(1), 57–89. https://doi.org/10.1016/0271-5309(82)90035-0

Wu, J., Yang, J., & Honda, T. (2010). Fitts' law holds for pointing movements under conditions of restricted visual feedback. *Human Movement Science, 29*(6), 882–892. https://doi.org/10.1016/j.humov.2010.03.009

Zhai, S. (2004). Characterizing computer input with Fitts' law parameters—The information and non-information aspects of pointing. *International Journal of Human-Computer Studies, 61*(6), 791–809. https://doi.org/10.1016/j.ijhcs.2004.09.006

Chapter 13
Behavioral and Neurobiological Foundations of Judgment Under Uncertainty

Mu-Chen Wang and Shih-Wei Wu

Living in a world full of uncertainty and surprises, the ability to estimate the probability of uncertain events is critical to the success of our plans, actions, and the quality of our decisions. Starting in the 1970s, cognitive psychologists led by Daniel Kahneman and Amos Tversky began their journey into studying *judgment under uncertainty*—how people estimate the probability of events. Their works and many others revealed the imperfect nature of our judgments and highlighted the use of heuristics when making probability estimates and predictions. Neurobiologists, unlike psychologists, did not begin their journey in studying probability judgments, although many learning, decision-making, and inference problems they investigated require estimating the probability of events. As a result, behavioral and neurobiological investigations on judgments under uncertainty overlap, but their findings on the principles and building blocks for judgment under uncertainty have not converged. In this chapter, we reviewed literature in both fields and discussed potential directions for future research that would lead to greater synthesis between the psychology and neurobiology of judgment under uncertainty.

13.1 Behavioral Literature

Since Kahneman and Tversky's seminal work in the early 1970s, psychological and behavioral research on how people estimate the probability of uncertain events has blossomed. The key insight from these studies indicated that people's probability

M.-C. Wang · S.-W. Wu (✉)
Institute of Neuroscience, National Yang Ming Chiao Tung University, Taipei, Taiwan
e-mail: swwu@nycu.edu.tw

S.-W. Wu
Brain Research Center, National Yang Ming Chiao Tung University, Taipei, Taiwan

D. H. Wu (ed.), *Cognitive and Neural Foundations of Chinese Reading*, Chinese Language Learning Sciences, https://doi.org/10.1007/978-981-96-6669-0_13

judgments do not conform to the principles of probability theory. Many reviews and books had already discussed the literature. Our attempt here was to focus on works that either examined Bayesian models for human judgments or studied belief updating in dynamic, changing environments. We chose these two topics because they are closely connected with current neuroscience investigations on reward-related learning and decision-making. Hence, our review specifically discussed the following four different lines of research: base rate neglect (13.1.1), base rate neglect vs. conservatism: prioritizing strength and weight of evidence (13.1.2), perceptual judgments and decision-making (13.1.3), and judgments in changing environments: reversal learning and change detection (13.1.4).

13.1.1 Base Rate Neglect

To illustrate base rate neglect, consider the following example from Kahneman and Tversky (1973):

> Jack is a 45-year-old man. He is married and has four children. He is generally conservative, careful, and ambitious. He shows no interest in political and social issues and spends most of his free time on his many hobbies which include home carpentry, sailing, and mathematical puzzles.

The subjects were told that "A panel of psychologists have interviewed and administered personality tests to 30 engineers and 70 lawyers, all successful in their respective fields. On the basis of this information, thumbnail descriptions of the 30 engineers and 70 lawyers have been written. You will find on your forms give descriptions, chosen at random from the 100 available descriptions. For each description, please indicate your probability that the person described is an engineer, on a scale from 0 to 100" (Kahneman & Tversky, 1973).

This is the classical lawyer-engineer problem. In Kahneman and Tversky (1973), half the subjects were informed that among the 100 people, 30 were engineers and 70 were lawyers. The other half were told that 70 were lawyers and 30 were engineers. Kahneman and Tversky (1973)found that people showed base rate neglect: subjects' probability judgments were insensitive to the changes in the base rate—the proportion of lawyers and engineers. Instead, their judgments seemed to primarily depend on how representative the thumbnail descriptions were to the stereotypical impression of engineer or lawyer.

A series of studies that followed reported that the degree of using base rate information is subjected to several factors. For instance, when the causal relevance of base rate information was given, people tended to use the base rate more (Ajzen, 1977; Bar-Hillel, 1980; Tversky, 1982). In another study, Fischhoff et al. (1979) found that when subjects experienced several different base rates (within-subject manipulation of base rate), they tended to use the base rate more in their probability judgments. Ginosar and Trope (1980) manipulated the level of consistency in the individuating information in addition to the base-rate manipulation, but nonetheless kept the between-subject design. When individuating information is less consistent with stereotypical impressions, such individuating information becomes less diagnostic of the person of interest. In this case, Ginosar and Trope found that subjects relied more heavily on the base rate in their probability judgments. In addition, when category membership was similar (electrical engineer and aeronautical engineer rather than engineer and lawyer), the effect of base rate on probability judgments was also more pronounced. Resolving several methodological concerns—including difficulty in controlling information presented as verbal descriptions, lack of revelation of the random selection process involved, and incentive compatibility—Grether (1980) found that people tend to underweight the base rate (Grether, 1980). However, they do not completely ignore the base rate, suggesting that the use of the word "neglect" was too strong and therefore inappropriate. It was also shown that, in studies that attempted to replicate the lawyer-engineering problem, people do change their probability judgments in the correct direction according to the base rate despite large variability across studies (Koehler, 1996). If the evidence pointed to the use of base rate to a certain extent, a related question would be: are people closer to completely ignoring the base rate or the normative Bayesian? This question was not addressed in the review by Koehler (1996), but was pointed out in Kahneman and Tversky (1973) that participants were closer to complete neglect of base rate than to the normative Bayesian. It is also important to point out that subjects in these early studies were asked to make only a few probability judgments, and therefore, it was not possible to characterize the use of base rate at the individual subject level. In summary, humans do take into account base rate information but tend to underweight it. However, it remains unclear, across studies and tasks, whether people are consistently closer to either complete neglect of base rate or normative Bayesian.

The normative Bayesian model also makes the predictions that the expected accuracy of a prediction based solely on the individuating information (e.g., the thumbnail description of the person of interest in Kahneman and Tversky) should impact the use of base rate. That is, when predictions (e.g., judging whether Jack is an engineer) are likely to be false, the Bayesian model predicts that judgments should rely more on the base rate. To test this, Kahneman and Tversky (1973) told two separate groups of subjects different expected accuracies (55% correct or 27% correct). They found that while expected accuracy did impact subjects' confidence judgments, subjects in the low-accuracy group did not move closer to the base rate in their judgments (the rank order assigned to different graduate-school fields by the subjects according to the descriptions they saw of a person).

Information weighting is the heart of most, if not all, judgment or inference problems. The weighting of base rate relative to likelihood (or individuating information), independent of either the base-rate or expected-accuracy manipulations described above, can also be investigated in the context of uncertainty associated with prior distribution and likelihood function. When the uncertainty in the prior distribution is large relative to the uncertainty of the likelihood function, the posterior distribution is close to the likelihood function, indicating that larger weight should be assigned to the likelihood (Fig. 13.1). In contrast, when prior uncertainty is small relative to likelihood uncertainty, larger weight should be assigned to the prior. Recent studies manipulating uncertainty in the prior and likelihood showed that actual weighting by human participants often moves in the correct direction: as relative uncertainty in prior increases, subjects relied more on the likelihood information (Ting et al., 2015; Vilares et al., 2012; Yang & Wu, 2020). Moreover, subjects did not change their weighting as much compared with the normative Bayesian (Ting et al., 2015; Yang & Wu, 2020). In Yang and Wu (2020), it was found that this pattern was primarily driven by the prior uncertainty, especially when there was little uncertainty in the prior distribution. When prior should be weighted more heavily due to its low uncertainty, subjects exhibited a tendency to assign larger weights to likelihood instead compared with the normative Bayesian. Together with early studies, these results pointed out that *insufficient adjustment* in the weighting of prior and likelihood information is a fundamental feature in human judgment under uncertainty.

Although early and recent studies shared the same goal of understanding how and how well humans combine prior and likelihood information, the empirical paradigms used are quite different. In early studies, subjects were provided with individuating information in the form of a personal sketch or verbal descriptions of an individual. Grether (1980) pointed out several methodological concerns, namely that it can be difficult to control information presented as verbal descriptions and that subjects were not informed about the random process being examined. In more recent studies, the

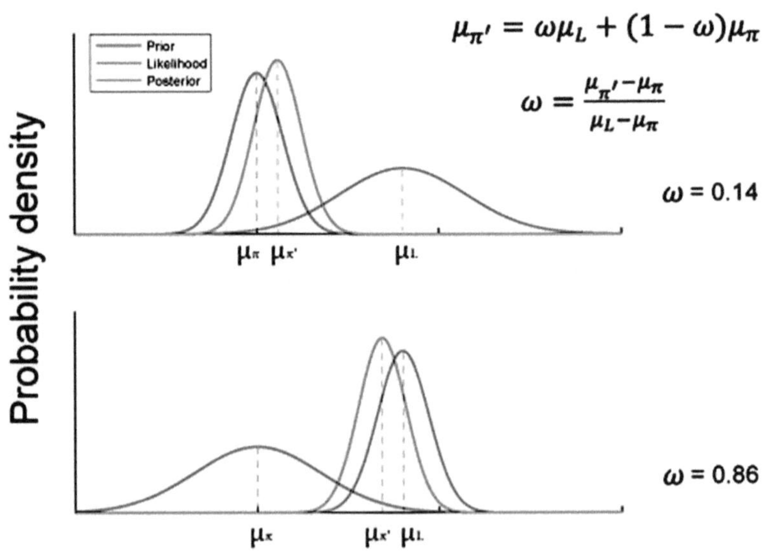

Fig. 13.1 Bayesian inference: examples. When the prior distribution π has a smaller variance than the likelihood function (L), the posterior distribution π' is closer to the prior distribution (upper graph). By contrast, when the likelihood function has a smaller variance than the prior distribution, the posterior distribution is closer to the prior distribution. The mean of the posterior distribution ($\mu_{\pi'}$) is the weighted average of the prior mean (μ_π) and mean of the likelihood function (μ_L). . The weight assigned to the likelihood relative to the prior (ω) reflects the variance of the prior distribution and likelihood function

paradigm shifted. For example, in Vilares et al. (2012), subjects had to guess the location of a hidden target on the computer screen. Prior distribution on target location was not explicitly stated and had to be learned through feedbacks after each trial. Likelihood information was colored dots sampled from a Gaussian distribution where the mean was the actual target location. Similar design principles can be traced by Kording and Wolpert (2004) and Tassinari et al. (2006). In other studies (Ting et al., 2015; Yang & Wu, 2020), prior distribution and likelihood function on reward probability (both as beta distribution) associated with visual stimuli were implemented and manipulated. In summary, these studies highlight two major departures from the Kahneman-Tversky paradigm. First, likelihood is directly drawn from the prior distribution, and samples were drawn from likelihood and visually presented to the subjects. Second, prior distributions are not explicitly stated and have to be learned through experience. It is obvious that differences in the degree of base rate neglect can be due to differences in task designs. However, it would be challenging to resolve task differences by coming up with an equivalent design between different tasks, for example, matching verbal descriptions with sampled dots on a computer screen for likelihood information. One potentially interesting direction is to have the same subjects perform different Bayesian-integration tasks so that we can compare the degree of similarity in base-rate neglect between them on a subject-by-subject level. The efforts to put forth such investigations remain to be seen.

13.1.2 Base Rate Neglect vs. Conservatism: Prioritizing the Strength Over the Weight of Evidence

While works from Kahneman and Tversky and others suggested that people underweight the base rate, studies prior to the base rate neglected findings that pointed out the opposite. Edwards (1968); (Ernst & Banks, 2002; Jazayeri & Shadlen, 2010; Knill & Richards, 1996; Kording & Wolpert, 2004; Stocker & Simoncelli, 2006; Tassinari et al., 2006)showed that people are overly conservative—they tend to rely more on the base rate and as a result do not update beliefs enough in the face of new evidence. One notable difference between these contradictory findings is on sample size: in the lawyer-engineer problem subjects saw descriptions of a single individual; in Edwards (1968), subjects were involved in a sequential-updating task where they were repeatedly shown new evidence.

To resolve these seemingly contradictory findings, Griffin and Tversky (1992) proposed that intuitive judgments are made based on the "strength" of evidence (e.g., warmth of a recommendation letter or proportion of heads in coin toss) and the "weight" of evidence (e.g., the creditability of the recommender, or number of coin tosses). They suggested that people tend to focus on the strength of evidence and show insufficient regard to the weight of evidence. This tendency causes people to overweight the strength of evidence when sample size is small (which leads to base rate neglect) but underweight the strength of evidence when sample size is large (which leads to conservatism). Therefore, insufficient regard to the weight of evidence can explain both base-rate neglect and conservatism.

The concept of strength and weight of evidence and the prioritizing of these different aspects of information are useful in thinking about how the brain works. One possibility is that different aspects of information, such as strength and weight of evidence, are being evaluated by different brain networks before they are being combined to produce judgments. In other words, different networks are specialized and selectively respond to either strength and weight. The prioritizing of strength over weight then operates at the across-network level, for example, with the strength-selective network dominating over the weight-selective network in the production of judgments (which may involve a third network). Alternatively, strength and weight might be represented by the same brain network. In this case, the prioritizing of strength over weight would operate at the level of neural response sensitivity within the network—greater response sensitivity to strength than to weight of evidence. We will return to this discussion in Sect. 2.5.

13.1.3 Perceptual Judgments and Decision-Making

While early studies on judgment under uncertainty found lack of evidence for Bayesian integration, more recent evidence coming from works on sensory perception and motor control showed quite the opposite results (Ernst & Banks, 2002; Jazayeri & Shadlen, 2010; Knill & Richards, 1996; Kording & Wolpert, 2004; Stocker & Simoncelli, 2006; Tassinari et al., 2006). It should be noted that perceptual psychologists have rooted interests in using statistical decision theory and signal detection theory to model perception (Green, 1960; Green & Swets, 1974; Swets et al., 1961; Tanner et al., 1954). This influenced the development of the "ideal observer analysis" (Geisler, 1989) that advocates for comparing human performance with performance predicted by some ideal or optimal model. Here, Bayesian model can be applied to define ideal/optimal performance, but it is not necessarily the only model for ideal performance. For example, in cue combination, the ideal performance is often defined by the principle of maximum likelihood estimation. Ernst and Banks (2002) showed that humans combine visual and haptic cues by taking into account their respective variance (noise) when judging the height of an object.

For testing whether perceptual judgments are Bayesian, a key challenge has been to specify and estimate the prior distribution. Unlike cognitive studies where prior such as base rate can be easily manipulated, humans in perceptual tasks may already have very strong priors that make it hard to experimentally manipulate the prior distributions. For example, in estimating the speed of moving stimuli, prior distributions favoring low speed have been used to explain biases in speed perception and motion-related illusion (Vintch & Gardner, 2014). Having said that, many perceptual experiments still elect to manipulate prior experimentally. For instance, with explicitly manipulated priors, Stocker and Simoncelli (2006) fitted a Bayesian model to the behavioral data (judging which moving stimulus has larger speed in a 2AFC task).

They found that the prior distribution for speed is positively skewed and has heavy tails at large speeds. This can explain why human speed judgments are often slower in low-contrast stimuli, although it did not directly test for Bayesian integration. In sensorimotor control, several studies in the mid-2000s manipulated the prior distributions and sample size to directly test whether motor control is Bayesian and found that human performance is very close to Bayesian (Kording & Wolpert, 2004; Tassinari et al., 2006). Jazayeri and Shadlen (2010) studied interval timing where human participants were asked to reproduce time intervals. Prior distributions on the time interval were uniform, and there were three different priors having different means. They found that subjects' performance can be well captured by the Bayesian model under one critical assumption. The uncertainty (standard deviation) in the likelihood function of subjects' interval timing perception increases linearly with the length of the sample time interval. For recent reviews on the perception literature, see Rahnev and Denison (2018) and Gardner (2019). In summary, it is possible that biases in judgments are domain-specific, with cognitive judgments like the ones studied by Kahneman and Tversky being more prone to biases than judgments in perceptual or motor tasks, although the use of heuristics in perceptual judgments had been reported (Laquitaine & Gardner, 2018). Future research needs to resolve differences in task design so as to directly compare performance between these different task modalities.

13.1.4 Judgments in Changing Environments: Reversal Learning and Change Detection

The judgments discussed in the previous sections are judgments in static environments where, despite uncertainty, the environmental state does not go through drastic changes. Indeed, a fundamental feature of the environment we live in is its dynamic nature, where shifts in environmental state (e.g., state of a pandemic, state of the stock market) are possible. The ability to detect changes is critical to perception, judgment, learning, planning, and decision-making. For instance, in studies of behavioral flexibility and learning, a reversal-learning paradigm was developed to specifically investigate how humans and animals detect and respond to changes in the reward or punishment contingencies in the environment. The reversal-learning paradigm has been extensively used as a test of cognitive flexibility in a host of neuropsychiatric disorders, including substance abuse, obsessive compulsive disorder, psychopathy, Parkinson's disease, and schizophrenia (Izquierdo et al., 2017). Lesion studies showed that there is dissociation between learning the reward contingencies and learning the change in reward contingencies. Patients with OFC lesions show deficits in reversal learning, although their performance in the initial stimulus-reinforcement learning remains intact (Fellows & Farah, 2003). Lesion of the OFC in monkeys showed deficits in performance after reversals (Rudebeck & Murray, 2008). However, more recent studies found that OFC lesions did not impair reversal learning and suggested that evidence for reversal-learning deficits in OFC lesions from previous studies was due to damage to the nearby fiber tracts but not to dysfunction in OFC (Rudebeck et al., 2013).

Since reward contingencies in the reversal-learning task is easy to learn, e.g., a stimulus would lead to a reward with certainty while the other stimulus would lead to no reward with certainty, subjects can develop simple strategies such as win-stay/lose-switch when reversal happens, which may limit the study of cognitive/behavioral flexibility. As a result, probabilistic versions of the reversal-learning task where subjects face between stimuli or actions carrying different probabilities of reward (e.g., 70% reward for one stimulus, 30% reward for the other), were developed. Probabilistic reversal learning task has been predominantly used in humans, and growing studies in monkeys and rodents are beginning to use it. A recent development has been to incorporate computational theories, primarily reinforcement-learning models and less so Bayesian models, into understanding behavior in the reversal learning task (Bartolo & Averbeck, 2020; Costa et al., 2015).

In studying change-point detection, several other interesting paradigms other than reversal learning had been developed (Behrens et al., 2007; Massey & Wu, 2005; McGuire et al., 2014; Nassar et al., 2010; Payzan-LeNestour & Bossaerts, 2011; Payzan-LeNestour et al., 2013; Yu & Dayan, 2005). Behrens et al. (2007) manipulated the reversal frequencies—referred to as volatility—in a reversal learning task such that for some periods of time the reward probability was stable (stable environment) and for others the reversals happened frequently. The Bayesian learner would weight the prior more than likelihood in stable environments but would do the opposite in unstable environments. In the reinforcement learning framework, this can be thought of as learning-rate adjustments: in a stable environment, the learning rate should be low. Conversely, in unstable environments (where recent experience is more predictive of the future) learning rate should be high. In the delta-rule reinforcement learning model, value is updated through prediction error—the difference between actual outcome and prediction—and the learning rate is effectively the weight assigned to prediction error. When the learning rate is large due to instability of the environment, prediction error would be weighted more and value updates would be more drastic. This, in effect, is similar to giving a larger weight to new evidence. Behrens et al. found that human participants did adjust the learning rate as predicted by the Bayesian framework.

The Behrens et al. paper inspired considerable interest, in the neuroscience circle, in studying learning and decision-making in dynamic environments where changes take place regularly. However, the task had many components and was complicated. First, the reward contingencies were probabilistic and therefore can be considered as a source of uncertainty. This source of uncertainty is referred to as "risk" by some authors, probably because in the decision-making literature, "decision-making under risk" refers to situations where decision-makers choose between lotteries—probability distributions on potential outcomes. Second, since subjects did not know the outcome of the unchosen option, they would have uncertainty about its reward probability (estimation uncertainty). In fact, subjects face an exploration/exploitation dilemma: if they choose the same option over and over (exploitation), she or he would not know what the alternative option might entail. By contrast, if subjects switch to choosing the option(s) she or he has less experience with (exploration), she or he would learn more about these options, which may or may not have better outcomes than the currently preferred one. Third, there was uncertainty about how frequently the environment changes—often referred to as reversal frequency, volatility, or unexpected uncertainty. Since these different sources of uncertainty can impact decision-making in dynamic environments, subsequent works further attempted to investigate their roles (McGuire et al., 2014; Nassar et al., 2010; Payzan-LeNestour & Bossaerts, 2011; Payzan-LeNestour et al., 2013). For example, Payzan-LeNestour and Bossaerts (2011) developed and fitted different Bayesian reinforcement learning models to the choice data and found that the Bayesian model with ambiguity penalty fitted data the best: participants tended not to explore enough—they did not do enough to reduce estimation uncertainty. In Nassar et al. (2010) and McGuire et al. (2014), they found that participants adjusted learning rate in response to change probability and relative uncertainty (uncertainty in one's belief about the current state of the environment), but the amount of adjustment was smaller compared with the Bayesian prediction.

In summary, from the reversal-learning paradigm that aims to study behavioral flexibility to tasks that systematically manipulated change probability to investigate how people learn in dynamic environments where changes happen, this line of work showed that both humans and animals respond to changes in the environment. Although the majority of these studies did not directly compare human performance with Bayesian models—as many of them involve fitting Bayesian models instead of comparing Bayesian model-predicted performance with actual performance—some studies did find that humans tend to underreact to changes compared with normative Bayesian (McGuire et al., 2014; Nassar et al., 2010).

While these studies provide important insights into the learning aspect of change detection, few have investigated the potential biases in change detection. One notable exception was the study by Massey and Wu (2005). Massey and Wu (2005) designed a regime-shift detection task and found that biases in change detection are affected by the stability of the environment and the noise in the signals that convey information about the current state (regime) of the environment. In particular, people tend to overreact to changes when receiving noisy signals but are nonetheless in stable environments where change is possible but unlikely. By contrast, underreactions were observed when signals are precise but the environments are unstable (Fig. 13.2A).

In their regime-shift detection task (Fig. 13.2B), subjects were told that there were two "regimes" and each was represented by an urn containing blue and red balls. The Blue regime had more blue balls, while the Red regime had more red balls. Each trial always started at the Red regime, but at any of the subsequent 10 periods during the trial could shift to the Blue regime. If a shift had taken place, there would be no further shifts throughout the rest of the trial. At each period during a trial, a ball was randomly drawn from the current regime—serving as a signal—and subjects had to provide an estimate of the probability that the current regime was the Blue regime, i.e., a probability estimate that the regime had shifted. At the beginning of each trial, subjects were explicitly told about the transition probability (how likely a shift would happen) and signal diagnosticity (ratio of the blue and red balls for the two regimes). For example, a transition probability of 0.01 indicates that at any

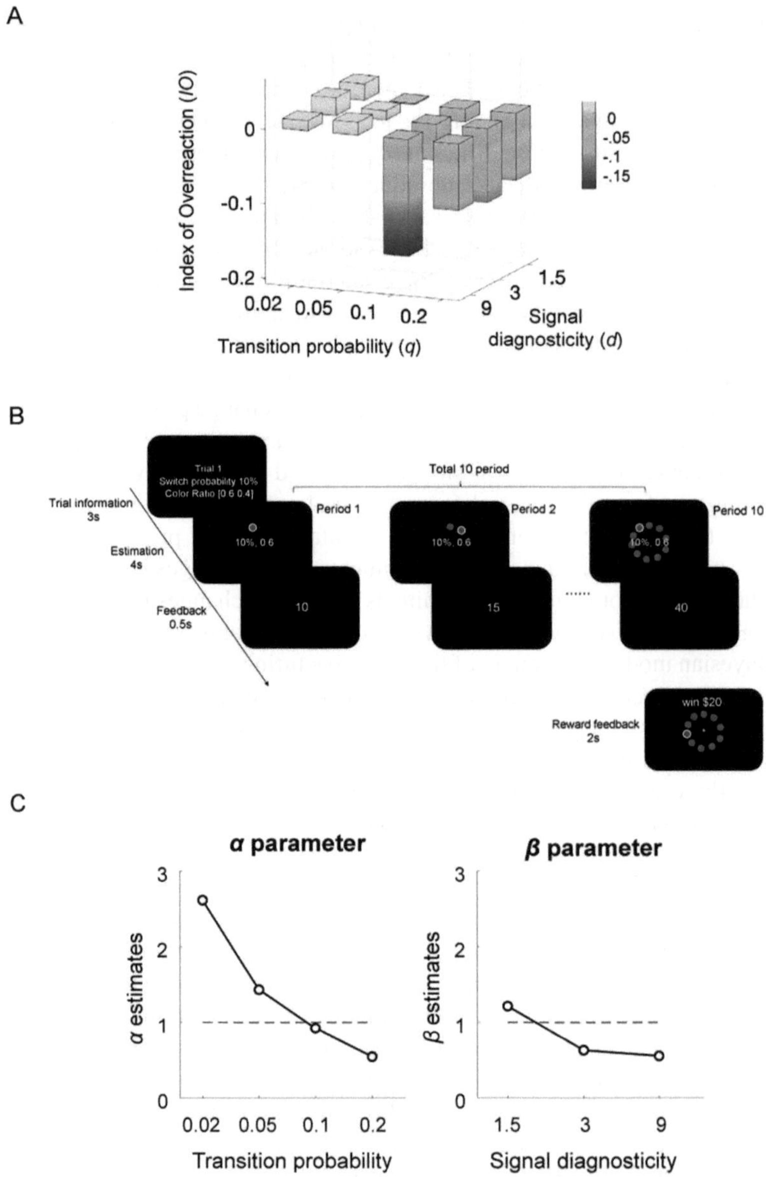

Fig. 13.2 Over- and underreactions to change. **A.** Over- and underreactions to change. Index of overreaction (IO) > 0 indicates overreaction; IO < 0 indicates underreactions. Data from Massey and Wu (2005) reanalyzed and plotted. **B.** Trial sequence of the regime-shift detection task. **C.** Parameter estimates of the system-neglect model. The α parameter models the distortion of transition probability, while the β parameter models the distortion of signal diagnosticity. Dashed lines represent the parameter value of the normative Bayesian decision-maker

period, there is a 1% chance that the regime would shift. For signal diagnosticity, a value of three indicates that blue balls are three times more than the red balls in the Blue regime, and in the Red regime, the red balls are three times more than the blue balls. The task was thus able to create different levels of environmental stability through manipulating transition probability and different levels of noise through manipulating signal diagnosticity.

The over- and underreactions observed in Massey and Wu (2005) were consistent with the "system-neglect" hypothesis: individual decision-makers respond primarily to the signals and secondarily to the system that generates the signals. In the context of the regime-shift task, the transition probability and signal diagnosticity are the *system parameters* that impact the production of the signals and are therefore referred to as the system parameters. They found that, compared with the normative Bayesian model (dashed lines in Fig. 13.2C), participants tend to be less sensitive to changes in the system parameters even though they did respond to these parameters in the correct direction (Fig. 13.2C). Here a lack of sensitivity can be characterized the steep slope of decreasing parameter estimates as system parameter values increase. By contrast, the normative Bayesian decision-maker responds to the system parameters as they are and has a slope of 0 (dashed lines). In short, Massey and Wu found that humans estimate changes in the environment by taking into account system parameters. However, we tend not to be as sensitive to these parameters as we should be. Such a lack of sensitivity leads us to under- and overreact to changes in predictable ways.

A major difference between the regime-shift detection task in Massey and Wu (2005) and the change-detection tasks mentioned before is the availability of information about the parameters relevant to change detection. In the reversal learning paradigm (including Behrens et al. who manipulated volatility) and other change-detection tasks (McGuire et al., 2014; Nassar et al., 2010; Payzan-LeNestour & Bossaerts, 2011; Payzan-LeNestour et al., 2013), subjects were not given information about the stability (volatility) of the environment or the diagnosticity of a signal. Instead, they had to acquire these different sources of information through experience. By contrast, in Massey and Wu (2005), subjects were explicitly given information about transition probability and signal diagnosticity. Would people react differently to these parameters depending on whether they are revealed explicitly or have to be learned through experience? Future studies need to investigate this question. One hypothesis is that acquiring system parameters through learning would introduce uncertainty in the estimation of these parameters. As a result, people would be less sensitive to system parameters in a learning task than in a descriptive task where information is 100% revealed to the subjects. To test this hypothesis, one way is to manipulate the uncertainty about the system parameters in the regime-shift detection task in a similar way as experimenters often do in studying ambiguity aversion (Becker & Brownson, 1964; Curley & Yates, 1985; Levy et al., 2010; Rustichini et al., 2005). For example, large uncertainty about transition probability can be introduced when subjects are told that it can be of any value between 0 and 0.1, whereas low uncertainty can be introduced when it is within a much smaller range (e.g., between

0.04 and 0.06). In both cases, the mean was 0.05, but the variance (uncertainty) was very different.

13.2 Neurobiological Literature

Unlike the behavioral literature, the neurobiological investigations have not put an emphasis on biases and errors in judgments. Although Bayesian modeling framework has had a strong impact—from understanding different cognitive functions to neural coding—the neural mechanisms that lead to judgment biases remain poorly understood. In the following sections, we began by reviewing literature using the Bayesian framework as guidelines to investigate how the brain represents prior and evaluates evidence (13.2.1). We then expanded the discussions to how the brain represents complex distributions (13.2.2 and 13.2.3) and belief updating in dynamic changing environments (13.2.4 and 13.2.5). Finally, we ended the chapter with our thoughts on the future directions that could foster a deeper understanding of human judgments and their neurobiological basis.

13.2.1 Neural Representations for Likelihood, Prior, and Posterior Information

In a seminal study, Roitman and Shadlen (2002) reported that single-unit activity in the lateral intraparietal area (area LIP) represents the accumulated sensory evidence over time for perceptual attributes (the direction of partially random moving dots) during perceptual decision-making. Many subsequent studies replicated this work. For reviews, see Gold and Shadlen (2007) and Shadlen and Kiani (2013). More recent evidence, however, pointed out step-function-like activity patterns in single-trial analysis in LIP during evidence accumulation, suggesting non-evidence-accumulation-like behavior (Latimer et al., 2015; Zoltowski et al., 2019).

In a typical perceptual decision task, sensory evidence in favor of a certain attribute (e.g., motion direction) cannot be quantitatively defined and manipulated. Although ramping activity in LIP neurons is consistent with evidence accumulation, these perceptual studies cannot directly examine whether LIP neurons represent likelihood ratio (or its logarithm). To investigate this issue, Yang and Shadlen (2007) designed a sequential weather prediction task where different visual cues associated with different likelihood of actions (left or right saccadic eye movements) being rewarded were sequentially presented. They showed that in monkeys, neuronal activity in area LIP represents the sum of log likelihood ratio over the sequence of observations. Subsequent works in humans using fMRI based on similar task design also found evidence for log likelihood ratio representations in the sensory cortex (Philiastides et al., 2010). In a value-based task where participants had to decide whether to buy or sell stocks, Gluth et al. (2012) found that anterior insula, pre-SMA (dorsomedial prefrontal cortex), and dlPFC tracked accumulated evidence for the likelihood that a stock would be profitable.

These studies, however, did not manipulate prior probabilities and therefore were not able to distinguish whether LIP neurons represent log likelihood ratio or log posterior odds due to the flat priors. Using the random-dot motion discrimination task, a perceptual decision paradigm widely used in the study of sensory evidence accumulation, Hanks et al. (2011) manipulated prior probability. They found that, at the behavioral level, the impact of prior probability depends on the quality of evidence. This is consistent with Bayesian inference. However, the key finding was that the impact of prior probability would increase over time during the evidence accumulation process, which is not entirely consistent with Bayesian inference. They further reported that LIP activity represents both the log likelihood ratio and the dynamic impact of prior (dynamic biased signal). Moreover, it appeared that the log likelihood ratio and the dynamic biased signal are additively combined to impact evidence accumulation in the decision process.

Although the model framework proposed by Hanks et al. (2011) was not Bayesian, the idea that the impact of prior probability increases over time as decision-makers are collecting evidence is similar to the idea of system neglect (see Sect. 1.4 in this chapter) where people respond primarily to the signals (evidence) and secondarily to the system that generates the signals (e.g., prior probability) (Massey & Wu, 2005). One possible neural implementation for system neglect thus might engage two distinct brain networks, with one specialized in evaluating evidence (including accumulating evidence over time when different pieces of evidence are presented sequentially) and the other in providing dynamic prior or base rate signals to influence evidence evaluation and subsequent decisions. The candidate network for evidence evaluation might be the frontoparietal network that includes intraparietal sulcus (IPS), previously shown to represent log likelihood ratio in monkeys (Yang & Shadlen, 2007). In humans, the frontoparietal network had shown to represent task difficulty as a result of noisy sensory evidence in perceptual judgment tasks (Heekeren et al., 2004).

More recently, a growing number of fMRI studies began investigating neural representations for prior, likelihood, and prior-likelihood integration. Vilares et al. (2012) found that uncertainty (standard deviation) of the prior distribution was represented in the putamen, amygdala, insula, and OFC. Among them, both OFC and putamen were found to track individual subjects' sensitivity to changes in prior uncertainty, while putamen was also found to correlate with individual subjects' weight given to likelihood relative to prior. By contrast, likelihood uncertainty was represented in the occipital cortex (superior occipital cortex extending to calcarine sulcus and lingual gyrus). d'Acremont et al. (2013) found that event frequency and subjective posterior were represented in separate cortical regions: event frequency was represented in the angular gyrus, posterior cingulate cortex, dorsomedial prefrontal cortex. By contrast, the subjective posterior was represented in the inferior frontal gyrus. Yang and Wu (2020), in a lottery choice task where information about prior and likelihood uncertainty about reward probability was manipulated, found that mPFC and OFC tracked individual subjects' weight given to likelihood relative to prior, and that putamen represented individual subjects' overall tendency to underweight the base rate. In coding reward probability, Ting et al. (2015) showed that the medial prefrontal cortex (mPFC) represents the mean of the posterior distribution on reward probability. Subsequent analysis further showed information about the mean of prior and mode of likelihood function in mPFC activity, suggesting that mPFC is critically involved in combining prior and likelihood information about reward probability. Using representational similarity analysis, Chan et al. (2016) found that medial OFC represented the entire posterior probability distribution on latent causes. Together, these studies found diverse sets of brain regions representing prior, likelihood, and posterior information, as well as individual subjects' relative weighting of likelihood or prior information during the decision-making process. The diverse findings on brain regions and networks involved are partly due to differences in task design, some were more sensory-based judgments (Vilares et al., 2012), some were about reward (Ting et al., 2015; Yang & Wu, 2020), some involved sampling from an urn containing two possible outcomes (d'Acremont et al., 2013), and some were about inferring latent causes (Chan et al., 2016). However, despite these differences, these studies converged on the key role of medial prefrontal cortex and OFC in prior-likelihood integration, regardless of whether rewards were directly involved or associated with prior or likelihood information.

13.2.2 Distributional Representations in the Brain

One key aspect of Bayesian integration is the representation of the prior distribution and likelihood function. How does the brain represent these distributions? Does the brain simply track summary statistics such as mean and variance? Or does it represent the entire distributions? Note that to perform Bayesian integration, it is not necessary to represent the entire distribution of the prior and likelihood. Under certain assumptions, it is sufficient to do so with the mean and variance statistics. Despite this, there is growing interest in studying how the brain represents an entire distribution, both at the theoretical level (Beck et al., 2008; Jazayeri & Movshon, 2006; Ma et al., 2006; Pouget et al., 2013; Soltani & Wang, 2010) and empirical level (Chan et al., 2016; Dabney et al., 2020; Geurts et al., 2022; Li et al., 2021; van Bergen et al., 2015).

At the theoretical level, one dominant framework—probabilistic population coding—posits that probability distributions are represented by population activity (Beck et al., 2008; Jazayeri & Movshon, 2006; Ma et al., 2006; Pouget et al., 2013). In essence, population activity is the activity of a group of neurons and therefore is a vector (with length equal to the number of neurons in the population) of neural response to a stimulus or input. In probabilistic population coding, Bayesian framework is applied to define and characterize how population activity represents probability distributions. When a stimulus is presented to a study participant, it elicits a neural response. Assuming uncertainty in the response, this can be modeled by a conditional probability $P(x|\theta)$ where x is neural activity and θ the stimulus parameter(s) (e.g., the direction of motion of a moving stimulus). The definition of x is somewhat flexible: x can be a vector of responses from a single neuron to the repeated presentations of a stimulus, or a vector of responses from a group of neurons to a stimulus on a single trial. Given that there can be many possible values of θ, we can define the likelihood function $\mathcal{L}(\theta; x)$ that describes the probability distribution of x over possible values of θ. Recognizing this, we can estimate the likelihood function $\mathcal{L}(\theta; x)$ empirically and use it for different purposes, e.g., to predict θ based on the posterior distribution $\pi\prime(\theta; x)$. For example, in van Bergen et al. (2015) human subjects performed a visual judgment task in which they had to judge the orientation of a tilted grating stimulus. Since it had been known that neurons in early visual cortex are orientation selective, van Bergen et al. estimated $\mathcal{L}(\theta; x)$ where θ represents stimulus orientation and x is brain activity (BOLD response in fMRI) from V1 to V3. Assuming non-informative prior $\pi(\theta)$, they were able to obtain the posterior distribution $\pi\prime(\theta; x)$.. They found that the uncertainty (standard deviation) of the posterior distribution matches the variability in the subjects' behavior (orientation judgments), suggesting that uncertainty regarding a stimulus attribute (visual orientation) is represented by a probability distribution encoded in neural population activity, consistent with probabilistic population code (Beck et al., 2008; Ma et al., 2006). Using a similar approach, Geurts et al. (2022) further demonstrated that the decoded probability distribution (using neural activity to decode the posterior distribution) matches subjects' confidence in these perceptual judgments. Further, they used the uncertainty (standard deviation) of the decoded probability distribution as a predictor and found that activity in the dorsal anterior insula (dAI), dorsal anterior cingulate cortex (dACC), and left rostrolateral prefrontal cortex (rlPFC) correlated with the decoded uncertainty. These results suggested that downstream areas (dAI, dACC, and rlPFC) may readout information about sensory uncertainty from the visual cortex and use it to compute subjective confidence.

Although these studies elegantly demonstrated that uncertainty is present in population activity that represents the probability distribution of stimulus attributes or features and that neural activity may readout these neural uncertainty estimates for decision-making, the neural mechanisms that give rise to suboptimal or biased judgments—a key finding in the behavioral judgment literature—remain elusive. This of course depends largely on the domain and kind of judgments. Nonetheless, to investigate neural mechanisms for suboptimal inference, it is important to develop behavioral paradigms that robustly show suboptimal behaviors and can identify factors that contribute to those behaviors (Ma & Jazayeri, 2014). For example, Yang and Wu (2020) found that prior uncertainty (the variance of the prior distribution) about reward probability plays an important role in base rate neglect in value-based decision-making. In a lottery choice task, they found that when prior uncertainty was small, indicating that prior should be trusted, human subjects showed base rate neglect by significantly underweighting the prior. By contrast, when prior uncertainty was large, subjects showed near-Bayesian performance. Further, Yang and Wu showed that this prior-uncertainty effect on base rate neglect was not due to incorrect knowledge of prior uncertainty due to memory decay. At the neural level, they showed that activity in the OFC and mPFC parametrically tracked relative subjective weight of likelihood (individual subjects' weighting of likelihood relative to prior) across different levels of prior and likelihood uncertainty. This suggests that base rate neglect arises in part as a result of subjective-weight computations in the OFC and mPFC.

13.2.3 Distributional Coding of Value Predictions

Although it had been shown, at both theoretical and empirical levels, how the brain can represent distributions, less is known about distributional representations of rewards and in reinforcement learning (RL). Traditionally, neural mechanisms for RL had been dominated by mean-tracking algorithms (Schultz et al., 1997). It was not until recently that distributional coding of reward value had been considered. In a recent study, Dabney et al. (2020) reported that GABAergic neurons in the ventral tegmental area (VTA) represent the entire reward distributions in ways predicted by distributional temporal-difference (TD) learning algorithm recently developed in machine learning. These GABAergic neurons exhibit different degrees of optimism and as a result, have different reward predictions. Some neurons are more optimistic, predicting larger reward outcomes, while others were more pessimistic, predicting smaller reward outcomes. These different reward predictions were developed as a result of each neuron's unique asymmetric weighting of positive reward prediction error (RPE) and negative RPE. Optimistic neurons tend to weight more heavily positive RPE than negative RPE during the learning process, while pessimistic neurons do the opposite. Using the response profiles of these neurons, Dabney et al. were able to decode the entire reward distribution the animals experienced. Subsequent work also suggested that prediction error is scaled by the level of surprise that corresponds to the probability of different outcomes—when the probability of a reward outcome is smaller, receiving that outcome is more surprising than receiving the same outcome when its probability is bigger (Rothenhoefer et al., 2021). This finding suggested that RPE responses in dopamine neurons carry information about reward distributions.

Several questions remain. First, how would distributional coding impact learning and decision-making? Evidence from behavioral economics suggests that reward expectations would impact risk attitudes: when people expect larger variability, they tend to be risk-neutral. By contrast, we tend to be risk-averse when we expect little or no variability in the probability of reward outcomes (Sprenger, 2015). This suggests that there is a potential dissociation between coding an entire reward distribution and performing computations that give rise to risk attitudes in choice bahavior. Even though the brain may come to learn reward distributions, ultimately, it is not the reward distributions per se that determine risk attitudes. Rather, risk attitudes might be the consequence of other value-related computations that use information about reward distributions. The question is whether risk attitude is a direct consequence of coding an entire distribution—which had been suggested in economic models for reference-dependent preferences (Kőszegi & Rabin, 2006)—and if so, what the neural mechanisms that lead distributional codes to impact risk-attitudes through decision computations?

Second, is distributional coding an advantage or disadvantage in dynamic, changing environments when reward contingencies change? The empirical studies and theory work on distributional coding deal primarily with static environments where reward contingencies are fixed. How would a distributional coding mechanism respond to dynamic environments where the state of the world (e.g., reward contingencies) can dramatically shift from one to another?

To address this question, we first have to recognize a design feature in the reversal-learning paradigm. In a typical reversal learning task, there are only two reward outcomes: participants either receive a reward or no reward. In this case, because there are only two outcomes, distributional code does not seem to provide much advantage. Distributional code should be an advantage with increasing reward-outcome complexity, i.e., when there are multiple possible reward outcomes associated with an option. However, it is unclear how and how fast neurons would update the reward distribution of an option if it goes through drastic changes. In addition, it is also unclear how distributional TD responds to different shapes of reward distributions (e.g., uniform, unimodal, multimodal, etc.).

13.2.4 Change Detection, Expected Uncertainty, and Unexpected Uncertainty

Neuroscientists have always been interested in understanding how the brain adapts to changes in the environments. The notion of behavioral and cognitive flexibility in particular refer to how well humans and animals flexibly adjust to changes in the environments. In the domain of reward and reinforcement learning, the reversal-learning paradigm had been instrumental to the understanding of behavioral and cognitive flexibility (Izquierdo et al., 2017). A key advantage of this paradigm is that it can be readily implemented in different animal species, from rodents, monkeys, to humans. Hence, results from different species at both the behavioral and neural levels can be compared. As discussed in the previous section, in a reversal learning task the participants have to learn the reward contingencies, usually stimulus-outcome or response-outcome associations. The contingency usually reverses several times during an experiment. For example, two stimuli A and B initially carry 75% and 25% reward, respectively. At some point (usually after participants successfully learn the contingency), the contingency reverses such that A becomes 25% reward while B becomes a 75% reward.

The orbitofrontal cortex (OFC) has been shown to be heavily involved in reversal learning. In humans, neuroimaging studies showed increased OFC activity while performing a reversal learning task. Patients with OFC lesions often exhibit reversal learning deficits. In monkeys, both lesion and electrophysiology studies showed that OFC is critically involved in tracking changes in reward contingencies. Similar findings in rodents had also been reported. An emerging view on the role of OFC in reversal learning is that it represents expected outcomes of stimuli or actions, so that when a reversal occurs, OFC is involved in updating the expected outcomes. Other brain regions, such as the medial prefrontal cortex, striatum, and amygdala had also been found to affect reversal learning but to varying degrees. For a comprehensive review, see Izquierdo et al. (2017). Higher order change statistic that defines the frequency of change, such as the frequency of reversals, has been shown to be represented in the anterior cingulate cortex (Behrens et al., 2007) which can also be labeled as being part of the dorsomedial frontal cortex.

In characterizing dynamic environments where changes take place regularly, a useful distinction in modeling and identifying neural representations for the relevant reward and change statistics is the notion of expected and unexpected uncertainty (Soltani & Izquierdo, 2019). Broadly, we can define expected uncertainty as the sources of uncertainty expected by the decision-maker. For example, the variance of a reward distribution associated with a stimulus or action is considered a kind of expected uncertainty. In other words, decision-makers can expect that distributions with larger variance would yield greater uncertainty in the outcomes associated with a stimulus or action. In dynamic, changing environments, the change probability, if constant (as in most reversal learning tasks), can also be regarded as expected uncertainty. By contrast, unexpected uncertainty is any source of uncertainty that a decision-maker does not expect. This, of course, is highly subjective in nature. Hence, some authors (Soltani & Izquierdo, 2019) defined unexpected uncertainty as the subjective perceived changes in reward contingencies. In this case, adding uncertainty through implementing a probability distribution on change probability can induce unexpected uncertainty, but so can the very first time the subjects experience a contingency reversal in a changing but nonetheless predictable environments. Further, even with a constant change probability, a small change probability would likely introduce subjective perceived changes that are different from the response to a much larger change probability. Hence, there are many different sources that can impact unexpected uncertainty, which makes it harder to identify and track in the brain.

Works on the issue of expected and unexpected uncertainty in dynamic environments are growing. I discussed only two representative studies while leaving out many other papers. Using fMRI, Payzan-LeNestour and colleagues systematically examined the neurobiological foundation of expected and unexpected uncertainty. In a 6-arm-bandit tasks that systematically manipulated reward probability and change probability, they found that activity correlated negatively with unexpected uncertainty—at the time of reward feedback—in posterior cingulate cortex, postcentral gyrus, middle temporal gyrus, hippocampus, and posterior insula (Payzan-LeNestour et al., 2013). Payzan-LeNestour et al. also found that subjects' degree of uncertainty in the probability of reward (estimation uncertainty) was represented—at the time of reward-predicting cues—in intraparietal sulcus, middle occipital gyrus extending to the parahippocampal gyrus, striatum, middle frontal gyrus, and anterior cingulate cortex. The second study was by McGuire and colleagues (McGuire et al., 2014) examining computational factors that can impact learning rate in dynamic environments. The first computational factor is surprise, captured by change-point probability. The second factor is relative uncertainty about whether a change in environmental state had occurred (uncertainty in one's belief about the state of the environment). Participants can estimate relative uncertainty by analyzing the number of recent observations: as the number of observations attributable to the current environmental states increases, the relative uncertainty decreases. McGuire et al. found that frontoparietal network including dorsomedial frontal cortex, inferior frontal cortex, and the posterior parietal cortex dynamically track both change-point probability and relative uncertainty. Further, when comparing change-point probability with relative uncertainty, they found that visual cortex more significantly represented change-point probability, while the intraparietal sulcus (IPS) and ventromedial prefrontal cortex (vmPFC) more significantly tracked relative uncertainty about change. Together, these results suggested that dorsomedial frontal cortex is critical to belief updating in dynamic and changing environments. Both Payzan-LeNestour et al. and McGuire et al. showed that dorsomedial frontal cortex is involved in individual subjects' uncertainty about the current state of the world: in Payzan-LeNestour et al. the uncertainty was about the reward probability; in McGuire et al. the uncertainty was about whether a change in environmental state had occurred.

13.2.5 Over- and Underreactions to Change and Their Neurocomputational Substrates

Although recent studies were beginning to characterize, at the neurobiological level, belief updating in dynamic, changing environments, they did not in particular address how biases in belief updating or change detection might arise in the brain. To address this question, we recently adopted the well-established regime-shift detection task by Massey and Wu (2005) to investigate the neural mechanisms for over- and underreactions to change (Wang et al., 2023). Behaviorally, we replicated Massey and Wu and observed both over- and underreactions to change. More importantly, we were able to identify the factors contributing to these biases, namely, transition probability and signal diagnosticity. In a changing environment, transition probability defines the probability that the environmental state (also referred to as regime) shifts from one to another, and therefore, is a parameter that determines how stable an environment is. Signal diagnosticity defines how diagnostic a signal (sample) is to the true state of the world and is therefore a parameter that can be used to judge the strength of evidence in favor or against change. We found that overreactions were most commonly seen under small transition probability and poor signal diagnosticity. By contrast, underreactions were observed under large transition probability and great signal diagnosticity. These opposite behavioral tendencies—overreactions and underreactions—are therefore products of the interaction between transition probability and signal diagnosticity. In particular, we found that these biases were driven by a lack of sensitivity to transition probability and signal diagnosticity.

At the neural level, we found that change detection recruited two distinct brain networks: the ventromedial cortex-striatum network and a frontoparietal network. Ventromedial prefrontal cortex (vmPFC) and ventral striatum represented belief updating (Fig. 13.3A). Activity in this network correlated with subjects' probability estimates of change and belief updates—the signed difference between successive probability estimates. By contrast, the frontoparietal network correlated with the strength of change evidence and dynamically tracked transition probability (Fig. 13.3B). But are these networks responsible for over- and underreactions to change? Indeed, we found that sensitivity of vmPFC in response to transition probability tracked subjects' behaviorally measured sensitivity to transition probability, while sensitivity of frontoparietal network in response to signal diagnosticity tracked subjects' behaviorally measured sensitivity to signal diagnosticity (Fig. 13.3C). Given that the sensitivity to system parameters or lack thereof explains under- and overreactions to change, our results suggest that these behaviorally observed biases arise from vmPFC and frontoparietal network in how they respond to changes in the system parameters.

13.3 Future Directions

Behavioral research on judgment under uncertainty had traditionally focused on biases and errors in judgments, as seminal work from Kahneman and Tversky in the 1970s showed that people's probability judgments consistently deviate from normative Bayesian principles. Neurobiological research, by contrast, had not put an emphasis on studying judgment biases. Despite such differences, both fields extensively adopted Bayesian framework to understand judgment and decision-making. Recently, some neural studies pointed out that biases can arise from sensitivity in brain networks in response to environmental factors that should be considered when making probability judgments (see Sect. 2.5 in this chapter). For future search, we identify four important directions that can lead to greater synthesis between the two fields:

1. Characterize computational and parameter selectivity

How can we better understand judgment biases at the neurobiological level? Evidence emerging from recent studies suggested that computational selectivity and parameter selectivity are two organizing principles for judgment under uncertainty (Geurts et al., 2022; Vilares et al., 2012; Wang et al., 2023). Different brain networks are organized by the type of computations they specialized—for example, in evaluating the strength of sensory evidence—and by the environmental parameters or variables they selectively respond to (e.g., transition probability). In particular, Wang et al. (2023) suggested that over- and underreaction biases arise from sensitivity of

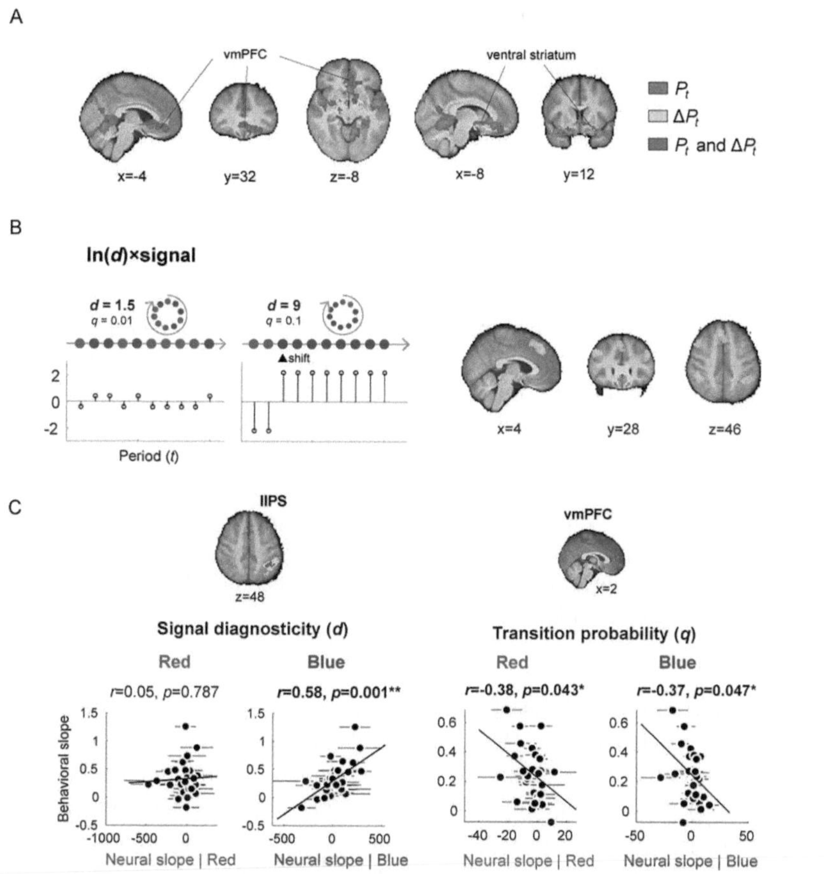

Fig. 13.3 Neural mechanisms for over- and underreactions to change. **A.** Belief revision. Brain regions that correlated with subjects' probability estimates (P_t) and the difference in P_t between successive periods (ΔP_t). Ventromedial prefrontal cortex (vmPFC) and ventral striatum activity correlated with both P_t and ΔP_t. **B.** Strength of change evidence. Activity in a frontoparietal network including the dorsomedial prefrontal cortex, inferior frontal gyrus, and intraparietal sulcus (IPS) correlated with the strength of change evidence, $\ln(d) \times$ *signal*—the interaction between signal diagnosticity *(d)* and signal. **C.** Network-level parameter selectivity. The frontoparietal network (here we use left IPS as example) represented individual differences in the sensitivity to signal diagnosticity, while the vmPFC correlated with individual differences in sensitivity to transition probability

brain networks in responding to environmental parameters critical to interpreting the signals we receive from the environments. It would be important for future investigations to further characterize how these two organizing principles lead to judgment biases. For example, are there ways that we can impact neural sensitivity at the individual subjects' level? How would changes in neural sensitivity to environmental parameters impact the behavioral biases? These questions remain open and deserve attention in future studies.

2. Develop neurobiologically based models for judgment under uncertainty

We believe that developing neurobiologically based models for judgment under uncertainty is crucial to better understand judgment and judgment biases at the neurobiological level. In studies of sensory perception and decision-making, neuroscientists are beginning to develop and use neurobiologically based models to explain and predict perceptual performance and decision-making behavior (for reviews Carandini & Heeger, 2012; Glimcher, 2022). It is widely observed across both cortex and subcortex that neuronal activity carries a "relative" code that can be modeled by divisive normalization (Carandini & Heeger, 2012). Divisive normalization has been used to describe various types of context-dependent decision-making behavior (Khaw et al., 2017; Louie et al., 2013; Webb et al., 2021; Zimmermann et al., 2018).

What would be the neurobiologically based models for judgment under uncertainty? It seems likely that divisive normalization can be involved. To better address this question, it is essential to identify the neural code for judgment under uncertainty. Currently, we lack electrophysiological findings across a wide array of judgment and inference-related tasks that can lead to the development of canonical computations for judgment under uncertainty. Hence, Electrophysiological signals at the single-unit and population level in both humans and animals must be obtained in different probability judgment and/or inference tasks.

3. Develop data-driven deep neural network models for judgment under uncertainty

Deep neural network models are excellent at learning functions without having to mathematically define them. The problem is that these models require a lot of data to learn. Recently, in an attempt to seek theoretical insights into decision-making, Peterson et al. (2021) trained deep network models to learn decision-making behavior from a big choice behavior dataset. Their approach was to exploit the flexibility of deep neural networks but use psychologically meaningful constraints—learning functions previously identified to be the key building blocks in decision-making (e.g., value function and probability weighting function in prospect theory). Because the neural nets are flexible, the researchers do not have to mathematically define those functions. They also do not face the constraints on the inputs to put in to learn functions. For example, Peterson et al. (2021) found that when inputs to the neural net include reward outcomes from both options to learn the value function in prospect theory, this resulted in a dramatic improvement in the model performance. This provided insight into the context-dependent nature encoding the value of potential outcomes previously not identified from behavioral research. In order to seek insights into judgment biases, an important direction would be to leverage the data-driven approach in deep neural network modeling to learn functions—for example, prior distribution and likelihood function—that are critical to derive probability judgments.

4. Identify the neural hierarchical architecture in probability judgments

Are computations for probability judgments supported by a hierarchical neural architecture? How might architecture relate to judgment biases? Understanding how cognitive functions are organized and supported by brain networks has always been a key interest to neuroscientists. Recent development in deep neural network modeling have inspired neuroscientists to understand hierarchical organization from the perspective of deep and interconnected layers of neural nets. In studying object perception, Yamins and colleagues (Yamins & DiCarlo, 2016; Yamins et al., 2014) used deep convolutional network to model the hierarchical architecture in the ventral visual pathway known for object perception. This has led to the discovery of a behavior-neural prediction match: the top hidden layer in the hierarchical convolutional neural network that best predicts behavioral performance on object recognition also predicts neural spiking data in inferotemporal cortex (IT cortex) best.

In the domain of probability judgments, like object perception, the brain also needs to first process the sensory information before such information can be read out by networks responsible for evaluating and combining information so as to compute judgments. However, in the study of judgment and decision-making, less is known about their hierarchical architecture. We do not even know for sure whether judgment computations are supported by hierarchically organized architecture. But what would be the advantage of knowing the architecture? One advantage is to identify the source of biases in the processing hierarchy—at what stage(s) in the hierarchy do biases emerge. And how might a deep convolutional network model help? Indeed, convolutional neural networks might not be appropriate for computing probability estimates. For object perception, convolution network is known to perform well with simple constraints—filtering, thresholding, pooling, and normalization—rooted in neurobiological principles. But these constraints would probably not work well to produce judgments. Therefore, constructing deep network models that include different sets of constraints seems necessary. For example, during the visual information processing stage, one might implement filtering, thresholding, pooling, and normalization constraints. When visual information is being read out by higher level layers, one might implement these layers to learn functions such as prior distribution and likelihood functions for computing judgments. In summary, developing and testing these hierarchically organized deep neural network models with neurobiological and psychological constraints has the potential to generate new insights into how judgment biases arise in the brain.

References

Ajzen, I. (1977). Intuitive theories of events and the effects of base-rate information on prediction. *Journal of Personality and Social Psychology, 35*(5), 303–314. https://doi.org/10.1037/0022-3514.35.5.303

Bar-Hillel, M. (1980). The base-rate fallacy in probability judgments. *Acta Psychologica, 44*(3), 211–233.

Bartolo, R., & Averbeck, B. B. (2020). Prefrontal cortex predicts state switches during reversal learning. *Neuron, 106*(6), 1044–1054 e1044. https://doi.org/10.1016/j.neuron.2020.03.024

Beck, J. M., Ma, W. J., Kiani, R., Hanks, T., Churchland, A. K., Roitman, J., Shadlen, M. N., Latham, P. E., & Pouget, A. (2008). Probabilistic population codes for Bayesian decision making. *Neuron, 60*(6), 1142–1152. https://doi.org/10.1016/j.neuron.2008.09.021

Becker, S. W., & Brownson, F. O. (1964). What price ambiguity? Or the role of ambiguity in decision-making. *Journal of Political Economy, 72*(1), 62–73.

Behrens, T. E., Woolrich, M. W., Walton, M. E., & Rushworth, M. F. (2007). Learning the value of information in an uncertain world. *Nature Neuroscience, 10*(9), 1214–1221. https://doi.org/10.1038/nn1954

Carandini, M., & Heeger, D. J. (2012). Normalization as a canonical neural computation. *Nature Reviews Neuroscience, 13*(1), 51–62.

Chan, S. C., Niv, Y., & Norman, K. A. (2016). A Probability distribution over latent causes, in the orbitofrontal cortex. *Journal of Neuroscience, 36*(30), 7817–7828. https://doi.org/10.1523/JNEUROSCI.0659-16.2016

Costa, V. D., Tran, V. L., Turchi, J., & Averbeck, B. B. (2015). Reversal learning and dopamine: A bayesian perspective. *Journal of Neuroscience, 35*(6), 2407–2416. https://doi.org/10.1523/JNEUROSCI.1989-14.2015

Curley, S. P., & Yates, J. F. (1985). The center and range of the probability interval as factors affecting ambiguity preferences. *Organizational Behavior and Human Decision Processes, 36*(2), 273–287.

d'Acremont, M., Schultz, W., & Bossaerts, P. (2013). The human brain encodes event frequencies while forming subjective beliefs. *Journal of Neuroscience, 33*(26), 10887–10897. https://doi.org/10.1523/JNEUROSCI.5829-12.2013

Dabney, W., Kurth-Nelson, Z., Uchida, N., Starkweather, C. K., Hassabis, D., Munos, R., & Botvinick, M. (2020). A distributional code for value in dopamine-based reinforcement learning. *Nature, 577*(7792), 671–675. https://doi.org/10.1038/s41586-019-1924-6

Edwards, W. (1968). Conservatism in human information processing. *Formal representation of human judgment.*

Ernst, M. O., & Banks, M. S. (2002). Humans integrate visual and haptic information in a statistically optimal fashion. *Nature, 415*(6870), 429–433. https://doi.org/10.1038/415429a

Fellows, L. K., & Farah, M. J. (2003). Ventromedial frontal cortex mediates affective shifting in humans: Evidence from a reversal learning paradigm. *Brain, 126*(Pt 8), 1830–1837. https://doi.org/10.1093/brain/awg180

Fischhoff, B., Slovic, P., & Lichtenstein, S. (1979). Subjective sensitivity analysis. *Organizational Behavior and Human Performance, 23*(3), 339–359. https://doi.org/10.1016/0030-5073(79)90002-3

Gardner, J. L. (2019). Optimality and heuristics in perceptual neuroscience. *Nature Neuroscience, 22*(4), 514–523.

Geisler, W. S. (1989). Sequential ideal-observer analysis of visual discriminations. *Psychological Review, 96*(2), 267–314. https://doi.org/10.1037/0033-295x.96.2.267

Geurts, L. S., Cooke, J. R. H., van Bergen, R. S., & Jehee, J. F. M. (2022). Subjective confidence reflects representation of Bayesian probability in cortex. *Nature Human Behaviour, 6*(2), 294–305. https://doi.org/10.1038/s41562-021-01247-w

Ginosar, Z., & Trope, Y. (1980). The effects of base rates and individuating information on judgments about another person. *Journal of Experimental Social Psychology, 16*(3), 228–242.

Glimcher, P. W. (2022). Efficiently irrational: Deciphering the riddle of human choice. *Trends in Cognitive Sciences, 26*(8), 669–687.

Gluth, S., Rieskamp, J., & Buchel, C. (2012). Deciding when to decide: Time-variant sequential sampling models explain the emergence of value-based decisions in the human brain. *Journal of Neuroscience, 32*(31), 10686–10698. https://doi.org/10.1523/JNEUROSCI.0727-12.2012

Gold, J. I., & Shadlen, M. N. (2007). The neural basis of decision making. *Annual Review of Neuroscience, 30*, 535–574. https://doi.org/10.1146/annurev.neuro.29.051605.113038

Grether, D. M. (1980). Bayes rule as a descriptive model: The representativeness heuristic. *The Quarterly Journal of Economics, 95*(3), 537–557.

Griffin, D., & Tversky, A. (1992). The weighing of evidence and the determinants of confidence. *Cognitive Psychology, 24*(3), 411–435.

Hanks, T. D., Mazurek, M. E., Kiani, R., Hopp, E., & Shadlen, M. N. (2011). Elapsed decision time affects the weighting of prior probability in a perceptual decision task. *Journal of Neuroscience, 31*(17), 6339–6352. https://doi.org/10.1523/JNEUROSCI.5613-10.2011

Heekeren, H. R., Marrett, S., Bandettini, P. A., & Ungerleider, L. G. (2004). A general mechanism for perceptual decision-making in the human brain. *Nature, 431*(7010), 859–862. https://doi.org/10.1038/nature02966

Izquierdo, A., Brigman, J. L., Radke, A. K., Rudebeck, P. H., & Holmes, A. (2017). The neural basis of reversal learning: An updated perspective. *Neuroscience, 345*, 12–26. https://doi.org/10.1016/j.neuroscience.2016.03.021

Jazayeri, M., & Movshon, J. A. (2006). Optimal representation of sensory information by neural populations. *Nature Neuroscience, 9*(5), 690–696. https://doi.org/10.1038/nn1691

Jazayeri, M., & Shadlen, M. N. (2010). Temporal context calibrates interval timing. *Nature Neuroscience, 13*(8), 1020–1026. https://doi.org/10.1038/nn.2590

Kahneman, D., & Tversky, A. (1973). On the psychology of prediction. *Psychological Review, 80*(4), 237.

Khaw, M. W., Glimcher, P. W., & Louie, K. (2017). Normalized value coding explains dynamic adaptation in the human valuation process. *Proc Natl Acad Sci U S A, 114*(48), 12696–12701. https://doi.org/10.1073/pnas.1715293114

Knill, D. C., & Richards, W. (1996). *Perception as Bayesian inference*. Cambridge University Press.

Koehler, J. J. (1996). The base rate fallacy reconsidered: Descriptive, normative, and methodological challenges. *Behavioral and Brain Sciences, 19*(1), 1–17.

Kording, K. P., & Wolpert, D. M. (2004). Bayesian integration in sensorimotor learning. *Nature, 427*(6971), 244–247. https://doi.org/10.1038/nature02169

Kőszegi, B., & Rabin, M. (2006). A model of reference-dependent preferences. *The Quarterly Journal of Economics, 121*(4), 1133–1165.

Laquitaine, S., & Gardner, J. L. (2018). A switching observer for human perceptual estimation. *Neuron, 97*(2), 462–474. e466.

Latimer, K. W., Yates, J. L., Meister, M. L., Huk, A. C., & Pillow, J. W. (2015). NEURONAL MODELING. Single-trial spike trains in parietal cortex reveal discrete steps during decision-making. *Science, 349*(6244), 184–187. https://doi.org/10.1126/science.aaa4056

Levy, I., Snell, J., Nelson, A. J., Rustichini, A., & Glimcher, P. W. (2010). Neural representation of subjective value under risk and ambiguity. *Journal of Neurophysiology, 103*(2), 1036–1047.

Li, H. H., Sprague, T. C., Yoo, A. H., Ma, W. J., & Curtis, C. E. (2021). Joint representation of working memory and uncertainty in human cortex. *Neuron, 109*(22), 3699–3712 e3696. https://doi.org/10.1016/j.neuron.2021.08.022

Louie, K., Khaw, M. W., & Glimcher, P. W. (2013). Normalization is a general neural mechanism for context-dependent decision making. *Proceedings of the National Academy of Sciences, 110*(15), 6139–6144.

Ma, W. J., Beck, J. M., Latham, P. E., & Pouget, A. (2006). Bayesian inference with probabilistic population codes. *Nature Neuroscience, 9*(11), 1432–1438. https://doi.org/10.1038/nn1790

Ma, W. J., & Jazayeri, M. (2014). Neural coding of uncertainty and probability. *Annual Review of Neuroscience, 37*, 205–220. https://doi.org/10.1146/annurev-neuro-071013-014017

Massey, C., & Wu, G. (2005). Detecting regime shifts: The causes of under-and overreaction. *Management Science, 51*(6), 932–947.

McGuire, J. T., Nassar, M. R., Gold, J. I., & Kable, J. W. (2014). Functionally dissociable influences on learning rate in a dynamic environment. *Neuron, 84*(4), 870–881. https://doi.org/10.1016/j.neuron.2014.10.013

Nassar, M. R., Wilson, R. C., Heasly, B., & Gold, J. I. (2010). An approximately Bayesian delta-rule model explains the dynamics of belief updating in a changing environment. *Journal of Neuroscience, 30*(37), 12366–12378. https://doi.org/10.1523/JNEUROSCI.0822-10.2010

Payzan-LeNestour, E., & Bossaerts, P. (2011). Risk, unexpected uncertainty, and estimation uncertainty: Bayesian learning in unstable settings. *PLoS Computational Biology, 7*(1), Article e1001048. https://doi.org/10.1371/journal.pcbi.1001048

Payzan-LeNestour, E., Dunne, S., Bossaerts, P., & O'Doherty, J. P. (2013). The neural representation of unexpected uncertainty during value-based decision making. *Neuron, 79*(1), 191–201. https://doi.org/10.1016/j.neuron.2013.04.037

Peterson, J. C., Bourgin, D. D., Agrawal, M., Reichman, D., & Griffiths, T. L. (2021). Using large-scale experiments and machine learning to discover theories of human decision-making. *Science, 372*(6547), 1209–1214.

Philiastides, M. G., Biele, G., & Heekeren, H. R. (2010). A mechanistic account of value computation in the human brain. *Proceedings of the National Academy of Sciences of the United States of America, 107*(20), 9430–9435. https://doi.org/10.1073/pnas.1001732107

Pouget, A., Beck, J. M., Ma, W. J., & Latham, P. E. (2013). Probabilistic brains: Knowns and unknowns. *Nature Neuroscience, 16*(9), 1170–1178. https://doi.org/10.1038/nn.3495

Rahnev, D., & Denison, R. N. (2018). Suboptimality in perceptual decision making. *The Behavioral and Brain Sciences, 41*, Article e223. https://doi.org/10.1017/S0140525X18000936

Roitman, J. D., & Shadlen, M. N. (2002). Response of neurons in the lateral intraparietal area during a combined visual discrimination reaction time task. *Journal of Neuroscience, 22*(21), 9475–9489. https://doi.org/10.1523/JNEUROSCI.22-21-09475.2002

Rothenhoefer, K. M., Hong, T., Alikaya, A., & Stauffer, W. R. (2021). Rare rewards amplify dopamine responses. *Nature Neuroscience, 24*(4), 465–469. https://doi.org/10.1038/s41593-021-00807-7

Rudebeck, P. H., & Murray, E. A. (2008). Amygdala and orbitofrontal cortex lesions differentially influence choices during object reversal learning. *Journal of Neuroscience, 28*(33), 8338–8343. https://doi.org/10.1523/JNEUROSCI.2272-08.2008

Rudebeck, P. H., Saunders, R. C., Prescott, A. T., Chau, L. S., & Murray, E. A. (2013). Prefrontal mechanisms of behavioral flexibility, emotion regulation and value updating. *Nature Neuroscience, 16*(8), 1140–1145. https://doi.org/10.1038/nn.3440

Rustichini, A., Dickhaut, J., Ghirardato, P., Smith, K., & Pardo, J. V. (2005). A brain imaging study of the choice procedure. *Games and Economic Behavior, 52*(2), 257–282.

Schultz, W., Dayan, P., & Montague, P. R. (1997). A neural substrate of prediction and reward. *Science, 275*(5306), 1593–1599.

Shadlen, M. N., & Kiani, R. (2013). Decision making as a window on cognition. *Neuron, 80*(3), 791–806.

Soltani, A., & Izquierdo, A. (2019). Adaptive learning under expected and unexpected uncertainty. *Nature Reviews Neuroscience, 20*(10), 635–644. https://doi.org/10.1038/s41583-019-0180-y

Soltani, A., & Wang, X. J. (2010). Synaptic computation underlying probabilistic inference. *Nature Neuroscience, 13*(1), 112–119. https://doi.org/10.1038/nn.2450

Sprenger, C. (2015). An endowment effect for risk: Experimental tests of stochastic reference points. *Journal of Political Economy, 123*(6), 1456–1499.

Stocker, A. A., & Simoncelli, E. P. (2006). Noise characteristics and prior expectations in human visual speed perception. *Nature Neuroscience, 9*(4), 578–585. https://doi.org/10.1038/nn1669

Tassinari, H., Hudson, T. E., & Landy, M. S. (2006). Combining priors and noisy visual cues in a rapid pointing task. *Journal of Neuroscience, 26*(40), 10154–10163. https://doi.org/10.1523/JNEUROSCI.2779-06.2006

Ting, C. C., Yu, C. C., Maloney, L. T., & Wu, S. W. (2015). Neural mechanisms for integrating prior knowledge and likelihood in value-based probabilistic inference. *Journal of Neuroscience, 35*(4), 1792–1805. https://doi.org/10.1523/JNEUROSCI.3161-14.2015

Tversky, A. K., Daniel. (1982). Judgments of and by representativeness. In D. K. P. S. A. Tversky (Ed.), *In Judgment under Uncertainty: Heuristics and Biases* (pp. 84–98.). Cambridge University Press.

van Bergen, R. S., Ma, W. J., Pratte, M. S., & Jehee, J. F. (2015). Sensory uncertainty decoded from visual cortex predicts behavior. *Nature Neuroscience, 18*(12), 1728–1730. https://doi.org/10.1038/nn.4150

Vilares, I., Howard, J. D., Fernandes, H. L., Gottfried, J. A., & Kording, K. P. (2012). Differential representations of prior and likelihood uncertainty in the human brain. *Current Biology, 22*(18), 1641–1648. https://doi.org/10.1016/j.cub.2012.07.010

Vintch, B., & Gardner, J. L. (2014). Cortical correlates of human motion perception biases. *Journal of Neuroscience, 34*(7), 2592–2604.

Wang, M.-C., Wu, G., & Wu, S.-W. (2023). System Neglect and the Neurocomputational Substrates for Over- and Underreactions to Change. *bioRxiv*. https://doi.org/10.1101/2023.05.30.542806

Webb, R., Glimcher, P. W., & Louie, K. (2021). The normalization of consumer valuations: Context-dependent preferences from neurobiological constraints. *Management Science, 67*(1), 93–125.

Yamins, D. L., & DiCarlo, J. J. (2016). Using goal-driven deep learning models to understand sensory cortex. *Nature Neuroscience, 19*(3), 356–365. https://doi.org/10.1038/nn.4244

Yamins, D. L., Hong, H., Cadieu, C. F., Solomon, E. A., Seibert, D., & DiCarlo, J. J. (2014). Performance-optimized hierarchical models predict neural responses in higher visual cortex. *Proc Natl Acad Sci U S A, 111*(23), 8619–8624. https://doi.org/10.1073/pnas.1403112111

Yang, T., & Shadlen, M. N. (2007). Probabilistic reasoning by neurons. *Nature, 447*(7148), 1075–1080. https://doi.org/10.1038/nature05852

Yang, Y. Y., & Wu, S. W. (2020). Base rate neglect and neural computations for subjective weight in decision under uncertainty. *Proc Natl Acad Sci U S A, 117*(29), 16908–16919. https://doi.org/10.1073/pnas.1912378117

Yu, A. J., & Dayan, P. (2005). Uncertainty, neuromodulation, and attention. *Neuron, 46*(4), 681–692.

Zimmermann, J., Glimcher, P. W., & Louie, K. (2018). Multiple timescales of normalized value coding underlie adaptive choice behavior. *Nature Communications, 9*(1), 3206. https://doi.org/10.1038/s41467-018-05507-8

Zoltowski, D. M., Latimer, K. W., Yates, J. L., Huk, A. C., & Pillow, J. W. (2019). Discrete stepping and nonlinear ramping dynamics underlie spiking responses of LIP neurons during decision-making. *Neuron, 102*(6), 1249–1258 e1210. https://doi.org/10.1016/j.neuron.2019.04.031

MIX
Papier aus verantwortungsvollen Quellen
Paper from responsible sources
FSC® C105338

If you have any concerns about our products,
you can contact us on
ProductSafety@springernature.com

In case Publisher is established outside the EU,
the EU authorized representative is:
Springer Nature Customer Service Center GmbH
Europaplatz 3, 69115 Heidelberg, Germany

Printed by Libri Plureos GmbH
in Hamburg, Germany